Protecting Your Fertility

The Danger Conventional Pest Control Poses to Fertility and Safe, Natural Alternatives Revealed.

BY
GABRIELA ROSA

ASSOCIATE PUBLISHING

Protecting Your Fertility

The Danger Conventional Pest Control Poses to Fertility and Safe, Natural Alternatives Revealed.

By Gabriela Rosa

ISBN-13: 978-0-9788010-8-3
ISBN-10: 0-9788010-8-3

Cover design by Embrace Design Solutions.
www.embracedesigns.com.au

Made in the United States of America

Get Your FREE Bonuses Today!

FREE Fertility Advice from 'The Bringer of Babies'

Leading natural fertility specialist, Gabriela Rosa (aka The Bringer of Babies) has a gift for you. As a thank you for purchasing this book get your FREE "Natural Fertility Booster" subscription and discover...

- Easy ways to comprehensively boost your fertility and conceive naturally, even for women over 40;
- Natural methods to dramatically increase your chances of creating a baby through assisted reproductive technologies such as IUI, IVF, GIFT or ICSI;
- Simple strategies to help you take home a healthier baby;
- How to prevent miscarriages.

You will also receive the FREE audio CD "11 Proven Steps To Create The Pregnancy You Desire And Take Home The Healthy Baby of Your Dreams" a total value of $397!

Claim your bonuses at

www.NaturalFertilityBoost.com

Be quick, this is a limited offer.

(Your free subscription code is: PYF)

Gabriela Rosa can be contacted via
www.BoostYourFertilityNow.com

Attention Gabriela Rosa

PO Box 2342

Bondi Junction NSW 1355

Australia

Dedication

To my wonderful patients who have taught me so much of what I know about fertility and life—I am honoured for your contribution. I also dedicate this work to you, the reader, for making my desire for greater learning, stronger than ever. I am infinitely grateful.

About the Author

Leading natural fertility specialist and naturopath Gabriela Rosa has gained international recognition as an expert in her field. Gabriela is the founder of Natural Fertility & Health Solutions—a multi-modality, integrative medicine centre based in Sydney, Australia. She dedicates herself to the management of women's health issues (from puberty to menopause), men's health and natural fertility treatment.

Gabriela is the author of *Eat your Way to Parenthood: The Diet Secrets of Highly Fertile Couples Revealed (2008), The Awful Truth About Cleaning Products and Fertility Revealed (2008) and Protecting Your Fertility—The Dangers Of Conventional Pest Control and Natural, Safe Alternatives Revealed (2008)* available from Amazon.com and all major bookstores worldwide. She has been featured in major Australian newspapers such as *The Daily Telegraph* and magazines such as *Woman's Day*. Gabriela has contributes to various radio programs in Australia and overseas, including Mix 106.5, 2UE, ABC and LA radio and she currently has a regular spot on Sydney radio.

In 2001, Gabriela created the highly successful **Natural Fertility Solution Program,** which she and her team run from her Sydney practice. With the aim of taking the program to couples all over the world, in 2008 Gabriela produced **The Natural Fertility Solution *Take-Home* Program**, which assists couples in overcoming fertility problems and provides them with the best possible chance of creating a healthy baby. The program is based on Gabriela's *11 Pillars of Fertility,* shown to dramatically increase a couples chances of a natural conception, while reducing the likelihood of miscarriage. The program has also shown to increase the odds for couples undergoing *in vitro* fertilization (IVF) procedures. Gabriela's program isn't just for couples with fertility problems—it's an essential toolkit for those who simply wish to prepare for the healthiest conception and baby.

Gabriela lives in Sydney with her husband.

Acknowledgements

I would like to thank and dedicate this book to all those people who throughout my life believed in me and for the lessons that made me grow beyond my wildest imagination. There are so many people to mention that this whole books would not be enough, however I would like to specially thank and acknowledge the following people:

Brook Canning, Vivienne Weinstock, Paul Doney and Paul Bayley—the incredible team at Natural Fertility & Health Solutions—without your support and friendship especially over the last year this project would probably still be unfinished! I am honoured to work with such amazing Integrative Medicine Practitioners as yourselves; To Carolina Silva, Gail Anne Waid, Carol Witt and Emma French without you this work would not be what it is today—I am eternally grateful for your support, your diligence and your tremendous ability to so beautifully transform my ideas into concrete reality; My family—mum, dad and Dani because you are my world and you continue to teach me so much about focus and determination; To my wonderful husband Maurice for your unwavering encouragement and delightful spirit—You are such a marvellous companion and I feel so fortunate to share my life with you; To Fofs because you are a massive cheering squad in such a tiny package; To Horton and Rishi, for bringing such fun and laughter into our lives! To Margaret and Aaron for being like my second parents—I love you both! To my new family Rosy, Michel, Marcel, and Michal thank you for helping me grow. And of course I couldn't forget to thank and appreciate my friends, teachers and mentors, present and past who even in the smallest ways continue to help, guide and encourage me to share the gift of optimum health and fertility with the world through their diverse and very special contributions: Dr Jim Ferry, Dr Jeff Jankelson, Francesca Naish, Jane Challinor, Mal Emery, Greg Owen and Wayne Pickstone.

Finally, my love to the future beneficiaries of this work and the gift of life—the babies I help bring into the world; through the loving intentions and consistent efforts of devoted and caring prospective parents who are touched by my message.

Foreword

To introduce the topic of toxic chemicals, pesticides and children's health I am going to first make the statement that "babies are being born polluted"

There is new research recently published which indicates that there is a linear correlation between tissue levels of pesticides and the incidence of insulin resistance. I will discuss this research to demonstrate an important point.

We first need to understand that insulin resistance is the new 'epidemic' metabolic disorder that has been creating the worldwide tsunami wave of obesity and diabetes and all the eventual complications of these two conditions.

It is commonly known that obesity with related diabetes can be traced back to genetics plus a high GI diet. It is simple, eat too many sweets, white bread, and donuts as a child, and end up fat and eventually diabetic.

However, we now have research that shows that just being born with high levels of pesticides in your bloodstream, will damage the hormonal system and cause insulin resistance. You could become obese and diabetic as a child, and sick and disabled as an adult, despite being brought up by parents who were careful to give you a healthy balanced diet, and despite a lifetime of 'doing the right thing'. Your parents just need to have been exposed to pesticides during their life and have accumulated tissue residues that can be passed on to you, as a baby in-utero and through breast-feeding. These chemicals then permanently damage sensitive developing systems.

Now, for the first time in history, children are being born with pre-existing levels of synthetic chemicals in their bodies. Of the hundred thousand plus chemicals involved, only a handful have ever been tested for toxicity effects on foetuses, and the handful that were tested, all demonstrated the ability to cause severe damage in some form. There was then a prolonged battle fought for years with the vested interests involved in these chemicals which blocked any regulation or restriction of their products.

Think tobacco; think asbestos or lead; and now think mercury, bisphenol-A, phthalates and antimony in children's clothing and bedding. Then wonder about the other 100,000 plus chemicals that have not even made it into the scientific spotlight or into the media.

A study was recently done in the US on the cord blood of newly born babies. They looked for a range of known toxins and found the majority were present AT BIRTH in the cord blood collected. Of these, 62% were known carcinogens, 75% were neuro-devolopmentally toxic, and many were known to produce birth defects or reproductive system damage.

Hence the statement "babies are being born polluted"

This counts as the biggest experiment ever conducted on humans, and one in which the results are unknown, but likely not to be pretty.

There is currently no system in place around the world to make manufacturers test chemicals for foetal and neo-natal impact. On the few occasions products are tested, it is on lab animals and on healthy young adults. We do not have a 'safety first' policy, and after these chemicals are released it may take decades before the damaging effects become known and action finally taken. By this stage, every living thing on earth will have measurable tissue levels of persistent organic pollutants (POPs).

In-utero development, which amazingly converts one single cell into a complex human being of 10 trillion cells, involves the DNA of the one original cell directing an impossibly complex production process. All the research indicates that this process is easily disrupted in-utero and in the neonate.

This probably explains the changes currently seen in paediatric diseases. Up to only a few generations ago, the challenge facing children was to survive the high incidence of deaths in childbirth and then to survive malnutrition-related and infection-related deaths in early childhood. Once past these barriers, the surviving children were generally fit and healthy.

Today this has all changed. In most countries infant mortality is low, and we are now faced with epidemics of new paediatric conditions that were not present only a few years ago.

The new face of childhood diseases is that of...

- Allergies, Epipens and nut-free schools
- Asthma
- ADHD and other learning and behavioural problems
- Autism
- Paediatric cancers
- Schools full of overweight, sickly children who, as adults, will have high incidences of infertility, obesity, diabetes, hypertension, heart disease, Alzheimer's, Parkinson's, infertility, cancer, mood disorders and so on.

There is no doubt that this is the world in which we now live.

If we change the way we do things, so that development, industry, economics and agriculture become 'sustainable' and eco-friendly, then perhaps we will give the next generation the best possible chance for a healthy body and brain.

Until then, Gabriela Rosa's advice is of critical importance. To give your child the best chance then parents (and prospective parents) need to make their bodies as healthy as possible, and the home needs to be as chemical-free as possible.

In toxicology, the rule always is that the best treatment is that of prevention. Much easier to keep chemicals out of our bodies, and therefore out of our babies, than to try and remove them once they have found their way in.

Dr Emmanuel Varipatis MB.BS. (UNSW)

-Integrative General Practitioner—YourHealth Manly, Sydney.

-Fellow of the Australasian College of Nutritional & Environmental Medicine (FACNEM)

-Board-Certifi ed Clinical Metal Toxicologist (IBCMT)

-Founding Board member MINDD Foundation

-Board-certifi ed with "Defeat Autism Now" of Autism research Institute, USA

An Additional Note from Dr Joel Bernstein

Gabriela needs to be congratulated for drawing attention to such an important topic. As well as opening our minds to areas of potential danger, she provides innovative and practical ways to make our lives safer. Unfortunately, human progress has led to the production of many toxic substances, some as by-products and others with dedicated uses such as cleaning products and pesticides.

Conventional science and medicine has for too long, been focused on trying to determine abnormal function at the cellular and biochemical level, and has largely ignored the bigger picture of prevention. This is an issue that relates to numerous synthetic substances from pesticides right through topharmaceuticals.

Considering our lack of knowledge, we must assume that avoiding exposure to potentially, or proven, harmful substances can only be beneficial to our general and reproductive health. The use of more natural products seems like a logical approach. Gabriela making us more aware of the risks of toxic chemicals, as well as providing us with a safer approach to pest control will add one more solid step towards achieving and maintaining healthy reproduction.

Dr Joel Bernstein | B.Sc. (Wits)., M.B. B.Ch (Rand)., FRCOG (London), FRANZCOG (Australia).

Fertility Specialist, Medical Director Fertility East IVF www.sydneyfertility.com.au

Contents

Dedication v

About the Author vii

Foreword ix

Introduction xix

The Truth about Commercial Pesticides—Your Health And Fertility 1

Symptoms of Pesticide Exposure 3

When to Call a Professional 5

What is Natural Pest Control? 7

How to Use This Book 8

What to Expect 8

Choosing Your Tools 9

A List of Basic Equipment and Ingredients 18

Plant Derivatives and Extracts 19

Important Know-How 27

Natural Pest Solutions To Safeguard Your Health And Fertility 29

Home Care 30

General Pest Control Recipes 32

The Pest Control Power of Plants 44
The Wonderful Mint Family 45
Pest-Specific Solutions 47
Ants 47
Bedbugs 67
Carpet Beetles 82
Cockroaches 88
Dust Mites 102
Fleas 110
Flies 124
Mosquitoes 131
Mice 136
Clothes Moths 144
Moths and Beetles Found in Food 150
Silverfish 157
Spiders 161
Body Care 167
Insect Bites and Stings 167
Insect Repellents 169
Lice 174

Scabies Mites 184

Ticks 187

Pet Care 191

Natural Pet Shampoos & Other Pet Care Recipes 192

Bedding and Outdoor Areas 202

Contacts and Resources 207

Introduction

A healthy, unpolluted and non-toxic environment is essential for optimum fertility. Everything in our macro and micro environments impacts on our health and fertility. Chemicals damage our health. They also have a huge impact on the Earth's delicate ecosystems and, not surprisingly, on the sperm and egg that will eventually become an embryo. That is why it is so important to make positive changes in order to overcome fertility problems and/or simply have the healthiest possible baby.

Everything you are exposed to, or do, for at least 120 days prior to a conception attempt will have huge repercussions on your ability to conceive, but even more importantly will have a dramatic impact on the health of your prospective child. Pesticides, herbicides and many other chemicals are not only toxic to our systems but they also imbalance fertility and hormonal levels in animals and humans. They mimic our own hormones, particularly oestrogen, and end up blocking receptors for other hormones such as testosterone. For women this is a problem because it enhances oestrogen dominance, leading to hormonal 'disproportion' and contributing to conditions such as endometriosis and fibroids. For men it disrupts testosterone balance and therefore healthy sperm production. Finally, it also disrupts embryonic development, particularly in relation to sexual development and reproductive function later in life. You can make a difference and safeguard your health, and the health of your prospective family, by changing your habits, namely dietary, lifestyle and even your cleaning and pest control of course!

From the outset, however, I must be clear that although every positive change is very important, each is but a single piece of the optimum fertility puzzle. Only together can they generate the necessary synergy to create profound and positive change. A whole person approach, taking all the pieces of the puzzle into consideration, is the essential way to restore and optimise natural fertility in men or women.

I have identified 11 key areas making up the platform I named the 11 Pillars of Fertility, which underpins my natural fertility treatment approach. In order to create the result many couples yearn for: The baby of their dream—these key areas require diligent and simultaneous implementation.

One of these important areas addresses the need to avoid toxic exposure to chemicals such as pesticides in order to truly optimise a couple's fertility as well as safeguard the health of a prospective child, hence the value of this book.

Fertility is not an isolated 'event'—it is intrinsically connected with emotional, whole body and even (believe it or not) spiritual health. The body and the spirit are constantly striving towards balance, good health, fertility, energy and happiness. The problem is that due to our hectic lifestyles, health compromises and poor daily choices, we often overlook the basic requirements for incredible health and optimum fertility.We only take care of ourselves when something breaks down rather than making daily choices that maintain our most precious possession to the absolute best of our ability.

Thankfully, given the right circumstances, the human body is capable of seemingly miraculous shifts. It takes sufficient time, energy, nutrients, lifestyle changes and other desirable conditions, but at any time, the choice is yours. You can choose to take the time to listen to your body, to learn from it and work with it.

There are empowering and endless possibilities, even in cases where all hope had previously been lost.

Frequently we hear of 'miracles' happening in other people's lives.

The key to making it happen is the belief in the possibility it can. True belief in these possibilities creates a change in attitude and mindset, which in turn makes you want to seek out further information and knowledge. This leads to a change in behaviour, which in turn creates different results and shapes your new reality.

Believe in possibilities, but do more—make the commitment to truly take the time to make friends with and nurture your body, as you would nurture your newborn baby—irrespective of any outcome, because it is only in doing so, that no matter what happens, you will create new possibilities for yourself and, who knows, hopefully even the baby of your dreams.

I have guided many couples using this approach and it has proven very successful for overcoming fertility problems, preventing miscarriages and increasing the chances of taking home a healthy baby—preventing miscarriages and malformation (even for older couples). However this is not its only application, it is also a vitally important approach for any prospective parents who simply wish to prepare for the healthiest possible conception and baby, giving a child the best possible start in life.

My patients are proof of this. I remember one couple who had had many complications and two stillbirths prior to seeing me. They implemented my recommendations and went on to have a healthy baby. They summed it up best when they said: "A healthy baby is 'a hope', not 'a promise'. We now understand that hope must be underpinned by specific actions and positive preparation and even then, it is not entirely in our hands, but at least we can now relax, knowing we've done our part and the very best we could means we can now relax." Helping you do your part, and the very best you can, towards fulfilling your dream of a healthy baby, is what this book is all about.

In this book you will learn to use safe, natural, chemical-free, non-toxic ingredients and methods for pest control. Besides making a big difference to your health and fertility, what you will learn to implement will also have a very positive effect on the environment and your hip pocket, because they are (in comparison) also extremely cost effective. You will learn how to protect your surroundings for you and your family's safety and you will also be doing the Earth a great favour!

I hope you enjoy this healthy, original way of controlling pests. Remember to have fun and experiment—your health, fertility and the environment will thank you!

With Love and Fertile Blessings,

Gabriela Rosa aka 'The Bringer of Babies'
BHSc, ND, Post Grad NFM, DBM, Dip Nut, MATMS, MNHAA
www.BoostYourFertilityNow.com

The Truth about Commercial Pesticides—Your Health And Fertility

Pesticides are used to eliminate cockroaches, flies, fleas, bedbugs, mosquitoes and other insects and pests, especially in public areas. They are poisonous chemicals that have been deliberately produced to be added to our environment and they continue to be used for two main reasons – they are big business and they at least appear to be successful in achieving their end result (although we generally end up with much more than bargained, despite only finding this out much later).

Pesticides have only one purpose, that of ridding us of the inconvenience of co-existing with insects although this can be done in a much healthier way to our general wellbeing and fertility. Pesticides are toxic and dangerous but have spread to every corner of the globe. The groundwater, surface water, soil, snow, rain, fog, and even the air we breathe are contaminated by them.

Residue from commonly available pesticides has been found to have harmful effects on the multitude of animal species on our planet. Just as disturbing is their harmful effect on mothers, the developing foetus, infants and children. Prospective mothers who are frequently exposed to the chemicals contained in the average pesticide product increase their risk of infertility, birth defects and miscarriage. Studies have also discovered connections between pesticide use and childhood leukaemia.

Despite the dangers of chemical pesticides (used for agricultural and/or household purposes), we know very little about their contents. Even the products you find on the shelves of your local garden centre do not legally have to be labelled with every ingredient they contain.

Even though we may not be aware of their presence, harmful chemicals surround us. They are used in our parks, the stores we visit, the fields where our food is grown, and even our homes. Did you know that antibacterial soaps and dishwashing liquids are listed by the Environmental Protection Agency (EPA) as pesticides?

The key is to protect yourself, through making the very best choices you can—not to use these and other chemicals, since dangers may also arise from pesticide use by neighbours, nearby plantations and local councils. To protect your health and fertility as well as that of your family, including the environment, education is essential. Understanding the risks and alternatives is the first place to start.

SYMPTOMS OF PESTICIDE EXPOSURE

(Certainly not a comprehensive list)

Allergies

Asthma

Behavioural abnormalities

Blurred vision

Central Nervous System disorders

Changes in heart rate

Chronic fatigue

Coma

Convulsions

Coordination loss

Cramps

Death (in extreme cases)

Diarrhoea

Dizziness

Elevated blood pressure

Fever

Flu-like symptoms

Genetic damage

Headaches

Hyperactivity

Immune deficiency disorders

Infertility (male and female)

Irritations to skin, eyes, nose and throat

Liver function impairment

Memory loss

Multiple chemical sensitivities

Muscle twitches and spasms

Nausea

Non-specific illnesses

Rapid heartbeat

Reproductive failures

Respiratory paralysis

Soreness of joints

Tightness in chest

Various cancers

Vomiting

Most troubling about toxic chemical pesticides are the unknown effects they have on our health and reproductive function. It may seem as though we know a lot about the dangers and health hazards presented by chemical pesticides, but this information is far from complete. There are many unknown health hazards that may take years of exposure before the effects are noticed and linked to their use.

While the large corporations that produce these dangerous chemicals do test each new product, they are often striving for results that will allow their product to be sold to consumers. The companies have spent millions of dollars creating the new product and do not want to lose their large investments because of a potential health hazard, so often the testing procedures for chemical pesticides and insecticides lack scientific rigour.

WHEN TO CALL A PROFESSIONAL

While I do believe it is possible to solve most household pest problems easily with natural products—the professionals do have their place.

Severe termite, fire ant or European wasp infestations can be overwhelming so there is no shame in calling in the pest busters. Perhaps you are allergic to fire ant bites? Termites can destroy a home. Would you risk your life or home to maintain a do-it-yourself job? I hope not.

The only rule to live by is 'All Natural, All the Time'. Do your research, and seek out a pest control company that guarantees to use only natural products. If the technician you speak with alludes to a 'less toxic' means of pest control consider what level of poison you deem acceptable for yourself and your family and choose wisely. Living in harmony with nature and establishing clear boundaries when bringing in others to aid you in your quest will pay dividends in health, fertility and wellbeing.

While professional pest control may be necessary in certain emergencies, try to avoid it, particularly during your 120 days of preparation prior to conception and definitely during pregnancy. If pest treatment has to be carried out, you could move into temporary accommodation until at least 6-12 weeks after any treatment—this is critical if you are pregnant. A second course of action might involve beginning your 120 days of preconception preparation after any pest control.

What is Natural Pest Control?

Natural pesticides and natural pest control methods can be used by humans without the risks to health and fertility normally associated with conventional (toxic) pesticides. As an example, most of us would consider pure vinegar to be harmless. We understand that if we drink enough vinegar we could be sick, or that if we get vinegar in our eyes it will sting, in other words, there is a fundamental understanding of how to safely handle and use vinegar. Using natural pesticides does not mean you can carelessly eat them, pour them into your eyes, squirt them all over your skin, or any such things. However, it does mean that if an accident occurs while you are using them and some spills on you, it won't mean a trip to hospital.

As the old adage goes, 'an ounce of prevention is worth a pound of cure.' This is particularly true in the case of pest control. One of the best ways to avoid an infestation or other pest-related problem is to stop it before it starts. This means if you spill a sticky drink wipe it up immediately before an army of ants finds it. Don't leave garbage sitting on your porch or in your garage, and clean up any damp papers, or any other damp moist areas that make a perfect breeding ground for unwanted insects.

Take an hour or so every few months to inspect your home for places where bugs could enter. Look for broken screens, gaps around doors and windows, and any other opening that looks inviting to pests. Once you have found them, take the time to make the necessary repairs or notify your landlord to ensure it gets done. Making these simple repairs can be the first step to preventing pests in your home.

HOW TO USE THIS BOOK

The primary purpose of this book is to introduce you to safe, natural ways to keep your home pest free. This will assist you in optimising your health and fertility prior to a conception attempt and during a healthy pregnancy. This book has been designed to assist you in creating balance between nature and your personal environment.

As you go through this book, you may find yourself skipping around a bit to find what you need in a particular situation. The table of contents and index will prove invaluable in your quest for quick answers to sudden flurries of pest activity.

You will find yourself returning to this book time and again, as new pest challenges present themselves from time to time. Like a good neighbour, this book stands at the ready, prepared to provide you with the information you need to confidently remain pest and pesticide free, naturally.

WHAT TO EXPECT

Everything in your environment will have an impact on your health. The aim is that through the use of this book that effect will be a good one. This book provides you with the necessary tools and information to clear your home of unwanted pests, and maintain it that way. It discusses various types of bugs and how to deal with them, but the best part is that if you have never before experienced a pesticide-free approach you are in for a pleasant surprise! Many non-specific ailments such as headaches, allergies and so on are likely to vanish as you bring harmony with nature into your personal habitat.

CHOOSING YOUR TOOLS

In this section, you will find more information on the primary essential oils recommended for use along traps, deterrents and the like. This is by no means meant to represent a full discussion of every essential oil available. It is, however, a useful basic list with which to begin assembling your anti-bug kit.

Essential Oil	Description
Angelica Root*	Spicy, forest scent, refreshing.
Anise	Deep, sugary fragrance, unique liquorice scent.
Balsam, Peru	Clean, refreshing, balsamic.
Basil**	Delightful, refreshing, with a unique liquorice scent. May also be used to repel bugs and treat bites
Bay	Fresh, exotic fragrance.
Bay Laurel	Clean, refreshing scent.
Benzoin	Luxurious, smooth, vanilla, light earthy scent.
Bergamot*	Refreshing citrus scent with a light hint of flowers.

Bergamot Mint*	Reminiscent of peppermint but lighter in fragrance. Slight hint of mint and citrus.
Boronia	Sweet and harmonious aroma.
Cajuput	Spicy, refreshing.
Cardamom	Sweet, velvety, exotic, forest-like.
Carrot Seed	Natural, warm, forest-like fragrance. This scent is not for everyone and does not smell at all like carrots.
Cedar wood**	Organic, forest-like, harmonious. Anexcellent natural insect repellent.
Chamomile**	Sweet, fresh, clear, vivid. Helpful for treating insect bites
(German or Roman)	
Cinnamon	Aromatic, rich, vivid, terrene (of the earth) exotic, spicy. Fragrance similar to ground cinnamon.
Citronella*	Refreshing, lemony, sweet. Known for its insect-repellent properties.

Clary Sage**	Vivid, natural, with a light fruity scent.
Clove Bud	Zesty, vivid, forest-like.
Coriander	Exotic, zesty, forest-like, sweet, with a hint of fruity aroma.
Cypress	Everglade, refreshing, natural.
Dill	Light, refreshing, slightly terrene, sweet.
Elemi	Refreshing, lemony, spicy.
Eucalyptus	Clear, vivid, refreshing. Also has medicinal attributes.
Eucalyptus, Lemon	Light, citrus, natural, slightly woody.
Eucalyptus Radiata	Aromatic, sweet, fresh.
Fennel	Lightly spicy fragrance, with a hint of liquorice.
Fir Needle	Crisp, refresh, terrene, woody.
Frankincense	Lightly spicy, terrene, crisp.

Galbanum	Crisp, terrene, woody, spicy, refreshing. Can be used to treat lice.
Geranium**	Sweet, crisp, floral, lightly fruity. May be used to treat lice
Geranium, Rose	Sweet, crisp, floral, light rose fragrance. Can be used to treat lice.
Ginger	Terrene, woody, spicy.
Grapefruit*	Lemon-like fragrance. Scent similar to fresh grapefruit but richer and more vivid.
Hyssop	Crisp, terrene, lightly floral.
Immortelle	Crisp, terrene, sweet.
Jasmine**	Rich, distinctive floral, spicy
Juniper Berry**	Fresh, terrene, herbal, nearly imperceptible fruity fragrance.
Kanuka	Terrene, herbal, sweet. Especially useful for skin irritations such as itching, ringworm and bug bites.

Lavender	Crisp, floral, herbal, rich velvety fragrance, refreshing. An essential oil for every home. It is indispensable for treating insect bites, itching, and as an insect repellent. It also has a myriad of other uses in the home.
Lavender**	Crisp, floral, slightly stronger fragrance than lavender. Can be used to soothe skin irritations and insect bites and to treat lice.
Lemon*	Fresh, crisp, aromatic, citrus. Fragrance similar to fresh lemons but richer and more vivid.
Lemongrass	Terrene, crisp, citrus. Especially useful as an insect repellent. Worth a spot in your 'bug be gone' kit!
Lime*	Crisp refreshing, sweet. Fragrance similar to fresh limes but richer and more vivid.
Linden Blossom	Light honey fragrance, rich, lemony, floral.

Mandarin	Sugary, citrus fragrance.
Manuka	
(New ZealandTea Tree)	Terrene, with a balsamic scent. Soothes irritated or itchy skin caused by rashes, insect bites, etc. A recommended essential oil for every home.
Marjoram	Dolce (sweet), herbal, forest-like, slightly antiseptic aroma.
Myrrh**	Terrene, warm, light balsamic fragrance. Helps soothe irritated or itchy skin caused by rashes, insect bites etc.
Neroli	Lemony, floral, sweet, vivid, crisp.
Niaouli	Terrene, rough, stale.
Nutmeg	Warm, spicy, velvety, sweet aroma. Spicy fragrance very like real nutmeg, but more vivid and rich.
Oakmoss**	Warm, vivid, terrene.
Orange, Bitter**	Orange citrus scent with a bitter edge.

Orange, Sweet**	Sweet orange aroma. Like fresh oranges but richer and more vivid.
Oregano	Spicy, herbal, brisk.
Parsley**	This is a natural abortifacient and should particularly be avoided during pregnancy.
Patchouli	Terrene, vibrant, lightly fruity. Useful insect repellent.
Pennyroyal**	This is a natural abortifacient and should =be avoided during pregnancy. Even topical contact with this oil can be hazardous.
Pepper, Black	Brisk, peppery, refreshing.
Peppermint**	Warm, fresh, crisp, minty, aroma very like peppermint candy, but richer and more vivid. Very good insect, pest rodent repellent properties. Also useful for treating scabies. An important essential oil for every home.
Petitgrain	Crisp, floral, lemony.

Pine, Scotch	Crisp, terrene, balsamic,.
Ravensara	Similar to eucalyptus, lightly sweet and fruity.
Rose*	*Rich floral, sweet and delicate.
Rosemary**	Aromatic, crisp, vivid scent, slightly sweet. Also has medicinal attributes.
Rosewood	Floral, sweet fragrance.
Sandalwood	Light, crisp, floral, woody, vivid.
Spearmint	Crisp, vibrant, lighter mint fragrance than peppermint. Especially useful for the treatment of scabies. Choose at least one oil from the mint family for your must-have list.
Spikenard	Terrene, stale.
Spruce	Crisp, refreshing, lightly fruity.
Tagetes	Delicate dolce fragrance, floral, with a light hint of fruit.

Tangerine	Bright, vivid, refreshing, citrus fragrance. Like fresh tangerines but with a richer and more vibrant aroma.
Tea Tree	Terrene, herbal, distinctive antiseptic aroma.
Tea Tree, Lemon	Slightly bitter, citrus aroma.
Thyme**	Aromatic, crisp, refreshing, herbal. Also has medicinal attributes. Helps soothe irritated or itchy skin caused by rashes, insect bites, lice, scabies etc.
Tobacco	Terrene, vibrant.
Tuberose	Intricate dolce fragrance, floral.
Vanilla	Creamy, smooth, sweet, vivid and velvety fresh. .
Vetiver	Terrene, velvety, herbal, exotic.
Violet Leaf	Light floral, terrene, vivid.
Yarrow	Bright, crisp, herbal.
Ylang Ylang	Delicate light floral fragrance, crisp, velvety.

*Use caution when handling.

**BEWARE: These essential oils are contra-indicated during pregnancy! Note: This list is intended as a guide, and while every care has been taken to make sure it is complete and up-to-date, new studies and information may change the safety warnings of these oils. If you have any questions as to the safety or toxicity of an essential oil, please consult your health care professional or herbalist before using.

A LIST OF BASIC EQUIPMENT AND INGREDIENTS

This is a list of the basic ingredients you will need to start making your own safe, natural pesticides. Most of them can be found inexpensively at your local health food or grocery store, or you may already have them at home. Take this book with you as a shopping list.

With the majority of suppliers now being online, you can place an order and have your supplies delivered to your door. For example, Dr. Bronner's castile soap can be ordered directly from the company's website. (For a complete list of suppliers, please see the resources section.)

- Castile soap
- Essential oils
- Boric acid
- Honey, sugar, jam

Ingredient Glossary

Here is a little more information about the ingredients you can use to ward off unwanted pests. Hopefully this section will shed some light on these too little used natural alternatives and will also become a source of information you can turn to whenever you need to know more about a natural alternative, its uses, and how to apply it.

PLANT DERIVATIVES AND EXTRACTS

DIATOMACEOUS EARTH

Where to buy:

You can find diatomaceous earth at garden centres and through natural pet care catalogues. Be careful to only purchase **natural** diatomaceous earth. Do not buy the type used for swimming pools.

Use:

Diatomaceous earth is applied directly as a pesticide.

About:

Diatomaceous earth is the fossilized remains of single-celled microscopic algae-type plants called diatoms.

Diatoms live in both salt and freshwater habitats, but are most commonly found in freshwater ponds.

During life, the plant secretes a delicate siliceous covering or shell. When the microscopic plant dies, its shell settles to the bottom. Over time, the shells of the deceased diatoms accumulate and form deposits called diatomaceous earth.

How it works:

When sprinkled on insects it causes damage to the outer shell or exoskeleton, making the insect dry out.

How to use:

It can be used in gardens as a natural insecticide and may also be used to repel bugs and prevent them from damaging food.

Another use for diatomaceous earth is to sprinkle a little in your vegetable seed packets if you will be storing them for an extended period. Alternatively, store unused seeds and seed packets in the refrigerator—seeds stored in the refrigerator will keep for 1 to 3 years and the cool temperature prevents any bugs from hatching and/or eating the seed.

BORIC ACID

Where to buy:

Supermarkets, discount stores, home and garden centres, pharmacies.

Use:

As an insecticide, especially for cockroaches.

About:

A naturally-occurring substance that is very useful for destroying cockroaches, termites, fire ants, fleas, and many other insects.

How it works:

Not fully understood but boric acid is believed to damage the stomach lining of cockroaches, possibly causing starvation.

How to use:

Can be used directly in powdered form for fleas and cockroaches, or mixed with sugar or jelly for ants.

LIQUID CASTILE SOAP (VEGETABLE OIL OR GLYCERINE-BASED)

Where to buy:

You can find liquid castile soap at health food stores or in the health food section of your regular grocery store, and through mail order companies. 'Dr. Bronner's' is one popular brand. The company has a 50-year history of making safe, natural products.

Use:

Soap is one of the oldest natural insecticides, having been employed for centuries.

About:

Most of these soaps are concentrated. Don't worry about buying the large size, thinking you will never use it all because liquid castile soap is the main ingredient of many of the recipes in this book.

Some plants are sensitive to soap and may suffer leaf burn if sprayed with a soap solution. Remember to test the solution on a small area before treating the entire plant.

How it works:

Soap eliminates insects by damaging cell membranes and causing dehydration and suffocation.

How to use:

Mix soap according to the recipes below, and then spray affected areas.

WASHING SODA

Where to buy:

Washing soda can be found in the laundry aisle of your supermarket.

Use:

It is one of the ingredients in natural pesticide recipes.

About:

Washing soda, also called sodium carbonate, is a white crystalline substance that can be extracted from plant ash, or produced synthetically from common salt. It is used in making soap, glass, paper etc. It is in the same family as baking soda but it isn't the same thing. It is processed differently and is much more caustic/alkaline, with a pH of 11. It doesn't give off harmful fumes but you should wear gloves when using it. You can usually find it in the laundry section of large supermarkets, often under the brand name 'Lectric Soda'.

How it works:

The high alkaline nature of this substance is irritating to a wide variety of insects.

How to use:

Used as an ingredient in the natural pesticide solutions found in this book.

BICARBONATE OF SODA (PH LEVEL ABOUT 8.0)

Where to buy:

Bicarbonate soda can be found in the baking aisle of the supermarket, or at discount stores, pharmacies, and bulk supply businesses.

Use:

Bicarbonate soda is helpful for preventing and eliminating mildew on plants. It has been used as a mildew repellent since the early 1900s. It is also a wonderful natural cleaning ingredient detailed in my other book The Awful Truth About Cleaning Products And Fertility Exposed.

About:

Historians believe that bicarbonate of soda has been in use since the time of the earliest civilisations. Be sure not to confuse bicarbonate of soda with washing soda. Bicarbonate of soda is a white crystalline substance, with a slightly alkaline taste. It is found in many mineral springs and can also be produced artificially. It is used in baking powders, and as a source of carbonic acid gas for soda water.

How to use:

To prevent crystallisation, which can occur when mixing bicarbonate of soda mixed with water, and to ensure an even coating on plants, mix bicarbonate of soda with a surfactant such as liquid castile soap? For best results, apply it weekly?

Exercise caution when mixing a bicarbonate solution. Follow the directions given in the recipes closely and do not add too much, which could cause leaf burn to some plants. Remember to test your solution on a small area before treating the entire plant.

Another bicarbonate, called potassium bicarbonate, is also very useful for treating mildew on grapes and roses and early blight on tomatoes and cucumbers.

Important Know-How

Here are a few things to keep in mind before you begin mixing your own natural pesticides:

Label, label, label

Always label your bottle before you begin. It will help keep your pesticides organised, make your pesticide treatment more efficient, and will save you time. In addition it will ensure there are not 'mix-ups'. If you are planning to store a pesticide solution for a while, put the name of the recipe, ingredients, and the date when made on the label. Labelling is also important in case of accidents. If the pesticides are labelled you will know exactly what is in them.

Re-using commercial chemical pesticide containers

Never re-use bottles from your old chemical pesticides. They can have harmful chemical residues that will contaminate natural pesticides. Even if the label makes the product sound harmless and you think the bottle might be okay, do not use it! Pesticide manufacturers often include harmful chemicals in their solutions, which they do not list on the label.

Never combine commercial chemical pesticides and home-made natural pesticides.

Never mix chemical pesticides and home-made pesticides. Without knowing the exact contents of each product and the chemical reactions that could happen between them, you could inadvertently make a combination that could release poisonous gases, or even cause an explosion.

Do not make your own mixtures

It is best not to try to create your own formulae, unless you really understand chemistry. Even if you mix apparently safe ingredients, you never know what reactions might occur.

Use caution when handling boric acid, liquid soaps and detergents

Although this may seem obvious it is important to reiterate caution when handling the ingredients you will be using to create your own natural pest control products. Be careful to avoid getting any of these ingredients in your eyes, as they can be very irritating. It is best to wear gloves whenever mixing products. Also avoid breathing in too much bicarbonate of soda or any other powdery substances—excessive amounts of particulate matter should not be entering your body (no matter how safe).

Natural Pest Solutions To Safeguard Your Health And Fertility

The recipes and pest solutions on the following pages have been organised into three main sections. In order to make your search for the right pest solution as quick and effortless as possible, a list of bugs and how to control them is alphabetically provided in each section.

The first section on 'Home Care' covers all the common pest problems found there, from one annoying fly, to ants in the sugar, to an infestation of fleas in the carpet. The second section is entitled 'Body Care' and it teaches you to naturally treat insect bites and stings—and most importantly how to prevent them by using some of the effective pest repellent recipes found there. This section also has detailed information on lice and nit treatments, shampoos and lotions. However please note: If you are allergic to insect bites or stings seek immediate medical attention. Do not attempt to treat the bite yourself.

Section three is dedicated to our furry friends, 'Pet Care'—providing great recipes for pet shampoos, flea powders, insect repellents and more. Pets, like humans, can have allergic reactions to some insect bites. If you suspect your pet has been bitten and is exhibiting signs of illness please seek immediate attention from your local veterinarian or pet clinic.

Home Care

Guide to Choosing Safe, Commercially Available Alternatives

There are many pesticide products on the market that claim to be natural but often contain inactive toxic ingredients that are potentially harmful to your environment and wellbeing. It is wise to exercise caution when purchasing these 'natural' pesticides, and if you have any concerns to call the manufacturer for more detailed information about the product's inactive ingredients. However, there are some commercial pest solutions that are safe organic- based alternatives.

Insect Powder

Pyrethrum powder (also known as Persian Powder or Insect Powder) has been used for centuries for the extermination of insects. This natural insecticide powder is made from the finely-ground flower heads of certain species of Chrysanthemum. It is important to check the label before buying any pyrethrum powder. Be sure it is pure organic pyrethrum powder, not synthetic, and that it does not contain any additives. This product should be used with caution because sensitive individuals may have allergic reactions. Pyrethrum powder can also cause allergic reactions in cats.

Pheromone Baits and Traps

Although pheromones are used in natural pest control to attract male insects and vermin into traps it may in fact also be a human hormonal disruptor, which will negatively impact on your fertility. Therefore, these are best avoided. In nature, pheromones are the scents released by insects (and other animals) to attract mates.

Ultrasonic Pest Control

There is some dispute as to whether or not ultrasonic pest control devices work.

Manufacturers claim the high-frequency sounds emitted by the machines effectively repel insects, birds, and pests such as mice. However, the majority of researchers have concluded that ultrasonic pest control machines do not live up to these claims. In addition these may also create undesirable electric magnetic radiation fields and thus are best avoided.

Havahart traps

These traps are especially designed to catch rodents, such as mice and rats. They can be baited with peanut butter or bacon. Once the rodent is captured, it can be released into the wild far from your home, without causing harm to the animal. These traps fold flat for storage and are popular because they effectively capture animals without killing or harming them.

GENERAL PEST CONTROL RECIPES

This section lists some natural methods for all-purpose pest control. These remedies are not species specific and therefore can be used on insects in general. Specific pests such as cockroaches, flies, fleas (and more) are also discussed in detail. All the recipes have been specially designed to fight bugs and pests using only natural methods and ingredients.

These general pest solutions help control and eliminate pests but regrettably, they also kill honey bees and other beneficial insects, like with any other pest control items it is important to exercise caution and use them sparingly. In addition, even though these recipes are natural, it is best to keep them out of the reach of pets and children, and use gloves and dust masks if necessary.

When working with essential oils:

Before using any essential oil, remember to read the warnings and safety information or check with a qualified aromatherapist or herbalist, as some essential oils may be inappropriate for use around pets, children, during pregnancy or by individuals who are especially sensitive.

Essentials oils may discolour wood or other surfaces, so test your chosen oil on a small unnoticeable area first and place cotton balls away from sensitive surfaces.

ORGANIC ESSENTIAL OIL INSECT REPELLENT

This will not only repel insects, but will also give your home a refreshing fragrance.

Ingredients:

Peppermint essential oil
Lavender essential oil
Citronella essential oil
Organic cotton balls

Yields: As required

Time to make: About 5 minutes
Shelf life: Indefinite
Storage: Separately in the original bottles or combined in glass bottle.

Method:

Apply a few drops of your chosen essential oil to a cotton ball.

How to use:

Place treated cotton balls around windows, doors, or by any place where you suspect insects are gaining access to your home.

BUGS AT BAY SACHETS

These delightfully-scented sachets will help keep insects at bay. Strategically placed by an open window, their refreshing scent will gently waft throughout the house.

Ingredients:

Fresh bay leaves
Cloves
Eucalyptus essential oil
Wedding favour bag or cheesecloth square
Ribbon, twine or string

Yields: As required

Time to make: About 5 minutes
Shelf life: Indefinite
Storage: As separate ingredients

Method:

In a glass bowl combine the fresh bay leaves, cloves and a few drops of eucalyptus oil and gently toss to mix. Place mixture in a cheesecloth square.

How to use:

Spoon the mixture into a small bag. A wedding favour bag works well, or you can make your own by sewing a small square from cotton cloth. Once the bag is filled, sew the edge closed.

Tip:

You can use ribbon or the draw string (if you used a wedding favour bag) to hang your finished sachet by an open door or window. This small fragrant sachet will help prevent insects from entering your home.

FLORAL INSECT-REPELLING MIST

This all-purpose mist will help prevent bugs from entering your home when sprayed around windows, foundations, doors, or any place where insects could enter the house.

Ingredients:

About 15 drops of one or two of the following insect-repelling essential oils:

Cedar essential oil
Citronella essential oil
Eucalyptus essential oil
Geranium essential oil
Rosemary essential oil
Tea tree essential oil
Thyme essential oil

Yields: 1 litre

Time to make: About 5 minutes
Shelf life: Indefinite
Storage: 1 litre misting or spray bottle

Method:

In a spray bottle, combine distilled water and about 15 drops of your chosen essential oil. Secure lid and shake well.

How to use:

Mist onto surfaces, such as door frames, window ledges, or around cracks and crevices, or anywhere you suspect bugs may enter. Shake well before each use.

ORGANIC HOME FUMIGATOR

Commercial toxic fumigators are among the worst offenders against a clean natural lifestyle. Not only do they infuse the air with harmful pollutants, they also leave entire rooms or homes quarantined until the dangerous toxins dissipate completely (which can take months).

This home-made natural fumigator is the perfect alternative. No rooms are left unusable with this toxin-free fumigator.

Ingredients:

500 ml distilled water
About 5 drops of one or two of the following insect-repelling essential oils:
Cedar essential oil
Citronella essential oil
Eucalyptus essential oil
Geranium essential oil
Rosemary essential oil
Tea tree essential oil
Thyme essential oil

Yields: As required

Time to make: About 15 minutes
Shelf life: Indefinite
Storage: Separately in the original bottles or combined in glass bottle.
Method:

In a small saucepan combine water and essential oil.

How to use:

Bring mixture to the boil, reduce heat and simmer. Allow the solution to gently simmer, being sure to remove the pan from the heat before all the water evaporates.

Tip:

You will need to keep a separate saucepan for fumigating, as traces of essential oils remaining in the pan may contaminate food. Because essential oils are highly concentrated they are not suitable for internal consumption.

Variations:

You may also use an essential oil diffuser (or oil burners) to release the essential oils into the air.

DIATOMACEOUS EARTH

Diatomaceous earth is an essential in your natural pest control arsenal. It works especially well on soft-bodied insects such as fleas, ants, spiders etc.

Ingredients:

Natural, not pool grade, diatomaceous earth.

Yields: As required
Time to make: Nil
Shelf life: Indefinite
Storage: Container with tightly-fitting lid

Method:

None required.

How to use:

Shake a fine layer of diatomaceous earth wherever you are having difficulties with pests. Thresholds, window seals, cracks and crevices are good places to apply diatomaceous earth.

For insects such as ants or fleas, which could be concealed in the fibres of your carpet, sprinkle diatomaceous earth over and around problem areas; wait for about three days and then vacuum. You can also gently brush the powder into the carpet.

Applications of diatomaceous earth can remain effective up to a year if kept dry.

Tip:

Diatomaceous earth is a fine powder that can damage computers, stereos and gaming consoles. To prevent potential damage it is wise to remove or cover electronic equipment before dusting the room with diatomaceous earth.

Wearing a dust mask is also recommended when using diatomaceous earth.

PLEASING PEPPERMINT SOAP MIST

Not only is peppermint one of the most soothing and refreshing scents, it is also a strong deterrent to invading insects.

Ingredients:

- **4 litres water**
- **2 toss castile soap**
- **10 drops of peppermint essential oil**

Yields: 4 litres

Time to make: About 5 minutes
Shelf life: Indefinite
Storage: Container with lid

Method:

Fill a bucket with water, add peppermint essential oil and castile soap. Mix well. Pour some of the solution into a spray bottle.

How to use:

Mist mixture along counter edges and backs, baseboards, under sinks and other areas where infestations may occur. Ants, fleas, flies and mice will all avoid areas misted with peppermint.

Variations:

Pepper has natural insecticidal attributes. To the above recipe, add one tsp of cayenne pepper. Adding cayenne pepper will reduce the shelf life of the solution to about three weeks.

GENERAL-PURPOSE SPICE INSECTICIDE

Rich aromatic spices combine to create an insecticide with a memorable aroma.

Ingredients:

1½ tsp. diatomaceous earth
1½ tsp. cayenne pepper
1½ tsp. garlic powder
1½ tsp. orange, lemon, or lime peels
1½ tsp. pyrethrum
1½ tsp. salt
1½ cups bay leaves
1½ cups peppermint leaves

Yields: About 1 cup

Time to make: About 25 minutes
Shelf life: Indefinite
Storage: Glass jar, such as those used in bottling and preserving.

Method:

Combine all the ingredients in a blender and grind to a fine powder. You can also use a pestle and mortar to grind the ingredients but this option is more time consuming.

How to use:

Sprinkle lightly at backs of bench tops, underneath cabinets, along baseboards and any other areas you suspect might be prone to infestation. Note: Do not rinse off or wipe up the powder.

GENERAL-PURPOSE PENNYROYAL MIST

Pennyroyal has been known since antiquity as an insecticide. The original French name meant 'a flea'. Today, in France just as in ancient Rome, pennyroyal is a good flea deterrent. .

Ingredients:

- **2 cups water**
- **1 tsp. Pennyroyal essential oil**
- **1 tsp. Eucalyptus essential oil**

Note: Pennyroyal is contraindicated in pregnancy. Consult your health care professional or herbalist before using. Keep out of the reach of children and pets to avoid accidental ingestion.

Yields: 500ml

Time to make: About 10 minutes
Shelf life: Indefinite
Storage: Spray bottle

Method:

Combine essential oils and water in a 500ml spray bottle. Secure lid and shake until well blended.

How to use:

Mist lightly at backs of bench tops, underneath cabinets, along baseboards and in other areas you suspect might be prone to infestation. Note: Do not rinse off or wipe up the misted solution.

Variations:

Woodsy Mist:

Substitute 2 tsp cedar essential oil for pennyroyal and eucalyptus oils in the above recipe.

Orange Mist:

For a delightful citrus scent, substitute 2 tsp orange essential oil for pennyroyal and eucalyptus oils in the above recipe.

SWEET AS HONEY INSECT TRAP

This solution is simple and quick to make! Bugs are lured to drink, become trapped in the water.

Ingredients:

- **¾ cup hot water**
- **3 tsp honey**

Yields: ¾ cup

Time to make: About 10 minutes
Shelf life: Indefinite
Storage: Glass jar with tightly-fitting lid.

Method:

Add honey to hot water and stir until dissolved.

How to use:

Pour honey and water solution into small shallow dishes and place anywhere where pests are a problem. This trap is especially effective for ants. Store any unused mixture in a glass jar with a tightly-fitting lid.

SOMETHING STICKY INSECT TRAP

This works in much the same way as the Sweet as Honey trap. Insects enter the jar, become stuck.

Ingredients:

- **1 cup hot water**
- **2 tbsp sugar**
- **2 tbsp molasses**
- **1 tsp active dry yeast**

Yields: 1cup

Time to make: About 10 minutes
Shelf life: Limited because fermentation may occur.
Storage: Glass jar with tightly-fitting lid.

Method:

Combine yeast with ¼ cup warm tap water and set aside. Heat the remaining ¾ cup water until boiling; remove from heat, add sugar and molasses and stir until dissolved. Combine yeast and sugar water and mix well.

How to use:

Pour mixture into a shallow glass jar or dish and place anywhere where pests are a problem.

THE PEST CONTROL POWER OF PLANTS

There is no better way to discourage pests than by welcoming a few beneficial plants into your home. Plants not only provide beauty and filter our air naturally to create a cleaner more breathable living environment. In addition, certain varieties and genera are anathema to bugs and rodents.

Planting these around the outside of your home, or bringing a few inside, will help deter pests. For example, grow an indoor herb garden on a sunny kitchen windowsill. Not only will you have fresh herbs ready at hand to use in your cooking, but the plants will help keep bugs and vermin out of the kitchen.

Some of the best herbs for general pest control:

Herb or Plant	Good Against
Basil	Flies, mosquitoes
Chamomile	All-purpose
Chives	All-purpose
Feverfew (contains pyrethrum)	All-purpose
Geraniums (any variety)	Mosquitoes, ticks
Hot Peppers (any variety)	All-purpose, especially cockroaches
Hyssop	Fabric and clothing moths
Lavender	Fabric and clothing moths
Lemon balm (citronella)	Mosquitoes
Mints (any variety)	All-purpose
Oregano	All-purpose
Pennyroyal*	Fleas, Attracts mosquitoes.

Peppermint	Cockroaches, mosquitoes
Rosemary	Fabric and clothing moths
Rue	All-purpose
Sage	All-purpose
Tarragon	All-purpose
Thyme	All-purpose
Wormwood	All-purpose

*Do not use when pregnant.

THE WONDERFUL MINT FAMILY

Mint can be grown near a doorway or window in an attempt to secure your home from the outside, before the pests even think of entering. Fewer insects around the entrance mean fewer insects in the house.

Mints are a hardy lot, and will claim any plot you allow them to spread in, so you may wish to consider pots of mint on your windowsill and along your porches. Mints adapt to most conditions, although they do like consistent watering.

Special Mint-Based Solutions

Create a mint barrier:

Mint leaves can be used as a barrier to discourage all manner of pesky prowlers. Chop the leaves to release their oils, and then scatter them in doorways, on window sills and around any cracks or crevices that are suspected of serving as entryways.

Mint Preventatives:

- Try tucking bags of peppermint tea into the corners of cupboards.
- A drop or two of spearmint, peppermint or pennyroyal essential oil on a fabric square (or on the teabags themselves) could be used similarly.
- The same essential oils can be added to your water when you wipe bench tops, shelves and floors.
- Ants seem to have a real love for moisture, so you may want to apply a few drops of the oils to pipes or taps where water collects.
- A few drops of essential oil can be applied to the hems of curtains that drift in the breeze.

In addition to mints, basil, citronella; cinnamon, lemongrass, lavender, and thyme essential oils are also effective in deterring a wide variety of insects.

Bundles of herbs hung in strategic places, can do more than add a touch of charm to your kitchen. Some pests are so sensitive to the herbs that they will avoid them. Tansy and common wormwood are particularly effective for keeping the pests away.

Pest-Specific Solutions

To tackle problem bugs we need recipes and solutions targeted specifically to them. The following solutions will help you combat everything from a solitary wandering bug to a whole infestation.

ANTS

It is appropriate that ants come first on our list of pests as they are one of the most common problem bugs for homeowners and picnickers alike.

Before setting about ridding ourselves of ants and other pests, it is wise, as the ancient Chinese warrior Sun Tzu observed, to 'Know your enemy'. Following is an overview of the general habits of ants.

Just like bees, ants have workers, as well as males and females. Males live only a short time, during which they may mate with the female 'Queen' ant.

Males are easily identified as they have wings, are smaller than females and die within about 24 hours. Queen ants have wings for a short time, and then shed them before beginning a new colony.

One hands-on method of eliminating new swarms is to watch for activity around established colonies, and then kill any queen ants you find. This requires timing as queens quickly begin digging new holes.

Ants also live in swarms, like bees, but they usually create hills of earth, with a labyrinth of chambers underneath the ground. In this contained system they store and keep their food supplies, raise their young and keep perfect order.

There are many different species of ant, with much diverse behaviour. 'White ants' or termites are not really ants, they belong to the Order Isoptera, while true ants belong to the Order Hymenoptera.

Ants come in as many different sizes and colours as there are different species. They range in colour from yellow to red to black, and in size from the very tiny (less than 1mm long) to the quite large at about 1.5 cm long. All ants have three body sections--the head, thorax, and abdomen.

Ants can be an irritation and danger to our wellbeing, because they bite humans and animals and may also carry disease. Their food sources include garbage and animal faeces, which harbour salmonella and other bacteria.

However, it is important to remember that ants, like bees, are also beneficial. They feed on flea larva, maggots, caterpillars, termites, adult fleas etc., reducing populations of hazardous insects naturally. Ants are really only a threat if they are a dangerous species, such as fire ants, hopper ants or inch ants.

You can purchase an illustrated insect and ant identification guide to help in distinguishing beneficial and dangerous species. As with all natural pest control methods, use caution and common sense when eliminating ants. Of course, if the health of your family is at stake, do not hesitate to contact a professional regarding identification or any other pest-related matter. Many people mistakenly believe that ants invade their home because it is dirty. This is not true. Ants will make themselves welcome wherever they find a food or water source, which can be anything from the sugar bowl to the rubbish bin.

Ants can be persistent and are not easily removed once they have chosen a location for a colony and settled in. Sometimes a new home will come with a decades-old ant hill, and ants in these colonies are very difficult to remove. However, with time, and consistent applications of the methods suggested in this book, even stubborn ant hills can be eliminated.

BASIC ANT CONTROL

Primary Ant Prevention

The main reason ants enter our homes is to find food or water so the best place to start is to remove these sources. Ant food sources can include rubbish bins, bits of food under cabinets or on bench tops. Even a small area of sticky cordial on the floor can be a source of food for ants. Begin by making sure all kitchen scraps are thrown away and bench tops are wiped clean. It is also a good idea to empty rubbish bins at the end of each day and take the garbage outside.

Ants are also attracted to compost bins, so make sure any compost containers are tightly sealed.

Garbage and scraps aren't the only food sources ants are attracted to; they will also enter sealed containers of sugar or cereal. Ants can even find their way into screw-top jars so it is a good idea to replace any screw-top storage canisters with air-tight ones.

One of the best ways to prevent insects from entering your home is to block any cracks or gaps leading to the outside. Look around your home for any places ants are likely to enter. Good places to check are around the foundations, electrical boxes, and door or window frames. Fill or caulk any holes, or sprinkle a natural ant deterrent such as cayenne pepper, mint, diatomaceous earth or boric acid around the area to keep out ants. For more ideas see the ant control recipe section.

Place ant deterrents such as mint, cayenne pepper, or cotton balls treated with citrus or tea-tree oil behind kitchen appliances, furniture, in drawers, shelves, wardrobes, and in the pantry. Turmeric powder or oil is another ant repellent to consider.

Don't be discouraged if one repellent doesn't work. Each insect reacts differently to natural barriers. You may need to try several ant deterrents before finding the one which works best for your ant species.

Three-Step Ant Control Program

There are so many ant species coexisting with us that it may seem impossible to find and implement the right control method for each. Happily, however, while there are many kinds of ants, they can all be controlled with one simple, easy, and effective program. Whether you suspect an indoor infestation, or can no longer enjoy a picnic on the lawn, this program will eliminate colonies both indoors and out, while reducing the chance of re-infestation. Even if you have tried other systems or remedies without success, I would urge you to try this comprehensive program.

Step 1: Inside treatment

This treatment consists of baiting the inside of your home. Baiting works especially well because ants live as a community, sharing food, so when worker ants carry poisoned food back to the colony from the bait it is distributed throughout the colony. This method is very effective and is also one of the only ways to kill the queen.

Choose a bait recipe and then place as many around your home as needed. You may need to experiment to find the recipe which best appeals to your ants. Ant tastes can differ from season to season, so you may need to change recipes if this occurs. Good locations for baits include behind furniture, behind large appliances, or anywhere you have noticed ant activity. Typically, indoor ant colonies are small and will seldom survive a comprehensive baiting programme.

Check the baits about once a week and refill if necessary. It is a good idea to keep a few baits around your home even if you are not currently experiencing any ant problems. These baits only need to be checked every four to six weeks to make sure they haven't evaporated.

Reminder:

Be sure to keep baits away from areas accessible to pets and children.

Notes:

*Baiting works best when it is the ants' only food source. Make sure all garbage is removed, bench tops and floors are free of crumbs and all food packages are sealed tightly.

*If you discover an infestation, it is important not to disturb the nest, and never to spray commercial pesticides on an ant nest. Not only are poisonous pesticides exceedingly harmful to you and your family's health, but spraying a nest with commercial pesticides cause the colony to divide, creating many new nests to remove.

Step 2: Outdoor treatment

The outdoor treatment focuses on treating the perimeter of your home. For best results you should treat the outside of your home on a monthly or bi-monthly basis, or after heavy rains to prevent new infestations.

Perimeter treatments are important because creating an outdoor barrier stops pests before they enter your home.

To create an outdoor barrier, draw a line around the foundations of your home using cayenne pepper or another ant barrier such as boric acid. Follow by treating any cracks in the foundations with boric acid or insect powder.

Step 3: Garden/Yard treatment

The final treatment tackles ants at their source by careful baiting of outdoor areas.

Begin your outdoor treatment by looking for any ant mounds. If you locate one, place a bait near it, or sprinkle a mixture of boric acid and sugar around it. (See the ant solution recipes section for more mound treatment ideas.)

Place baits behind shrubs and bushes, near garbage bin areas, and in any areas of known pest activity. Pests are often attracted to moist, damp, or dark areas, so consider placing baits in these areas.

Reminder:

Be sure to keep baits away from areas accessible to pets and children.

ANT SPECIES

There are about 3,000 species of ant in Australia and it is beyond the scope of this book to describe them all. Instead, here is an overview of some of the more common types of ants.

Carpenter Ants

Carpenter ants are so called because they make their homes in damp, moist or rotting wood. They shred the wood to make nests, but do not eat it. However, a serious, long term, infestation can destroy wood faster than termites.

They are generally very large in size and live in colonies with populations in the thousands. They are nocturnal, mainly foraging for food at night. This increases the chances of being bitten while you are sleeping, and carpenter ants have a very painful, stinging bite.

Workers typically forage alone, so if you notice a few stray workers wandering through your home this may be a sign of an infestation. Like other species of ant, they have winged queens and drones (males) which leave the nest, typically in spring, to start new colonies. If you notice large winged ants in your home it is very likely you have an infestation. If this happens, it is very important to make a thorough search to determine where the nest is located and its extent. Nests can be located almost anywhere, under leaky windows, in attics, behind kitchen appliances or cabinets, basements, walls, outdoors and anywhere there is wood in your home.

Fire Ants

Happily, for the average family in Australia aggressive, introduced fire ants should not be a problem. A progressive government eradication program, targeted at the only known infestations, succeeded in eliminating 99% of the fire ants as of February 2007.

However, if you suspect you have fire ants, do not touch or attempt to treat the nest yourself. The Queensland Department of Primary Industries and Fisheries requests that you immediately report any sightings to them--you can find additional information concerning the fire ant eradication program and its current status at: http://tinyurl.com/2jf5zg or http://en.wikipedia.org.

Pharaoh Ants

Ants are known to make nests in walls, ceilings, floors, and attics. One of the most common ants found in home interiors is the pharaoh ant. Pharaoh ants are considered a major nuisance. They are a tropical species, but will readily make their nests in any heated building. They are a particular problem in hospitals and nursing homes, as their very small size allows them to contaminate surgical wounds, IV drip systems, and sterile packaging, without being discovered. This ant has even been connected with some hospital infections.

Pharaoh ants are nocturnal and will be scouting for food while you are sleeping. If you discover a pharaoh ant nest in your home, it is important not to disturb it. One pharaoh ant colony can contain anywhere from two to 200 queen ants. Disturbing a pharaoh ant nest is likely to cause the colony to disperse, creating many new colonies. Unlike many ant species, neighbouring colonies do not fight.

You can treat the nest yourself using any of the recipes containing boric acid. Pharaoh ants are especially attracted to anything sweet, such as sugar, honey, and jams or jellies. However, they have been known to survive pest control treatments, re-emerging after the bait has been removed. If this happens, you may need to consult a professional.

ANT CONTROL RECIPES

ANT MOTEL

This easy-to-make ant trap is an effective way to control insects. It works especially well for capturing invading worker ants.

Ingredients:

1½ cups water
½ cup boric acid
½ cup sugar
Toilet tissue or cotton balls

Yields: 2 cups

Time to make: About 10 minutes
Shelf life: 1 to 2 weeks
Storage: Small glass jars with tightly-fitting lids. Small baby food jars work well.

Method:

Combine the water, boric acid and sugar in a glass jar. Shake well until the boric acid and sugar are mostly dissolved.

How to use:

Loosely stuff toilet tissue or cotton balls into a small jar, then add the sugar and boric acid solution to about an inch from the top. Tightly secure the lid. Using a hammer and nail carefully punch four to five holes into the lid. Place in areas where pests are a problem. Keep the solution fresh by replacing it weekly.

Note: Keep out of the reach of children and pets to avoid accidental ingestion.

QUEEN'S BANE POWDER

Getting to the queen of a colony is often a difficult task, as queen ants seldom leave their nests and often live many feet underground. However there is no better way to eliminate ant infestations than by addressing the queen.

Ingredients:

- **1 tbsp boric acid**
- **¼ cup powdered sugar**

Yields: ¼ cup

Time to make: About 5 minutes
Shelf life: Indefinite
Storage: Glass jar with tightly-fitting lid.

Method:

Combine boric acid and sugar in a glass jar, secure lid tightly and shake vigorously.

How to use:

Sprinkle the mixture around ant mounds, or in areas where you see ant activity.

Note: Keep out of the reach of children and pets to avoid accidental ingestion.

Tip:

Sensitivity to this remedy may vary between ant species and according to whether the colony has been exposed to this poison before. Because of these variables, you may need to adjust the sugar to boric acid ratio. The amount of boric acid in the sugar mixture should not be enough to immediately kill worker ants. You want them to collect the sugar and bring it back to feed the colony, which eliminates the entire nest. If the ants are dying where the power has been sprinkled, you will need to reduce the amount of boric acid in the recipe. On the other hand, if you do not notice any difference in activity after about three days, increase the amount of boric acid.

ANT JELLY

Ants are also attracted to jam, and in some cases prefer it to sugar.

Ingredients:

2 toss any jam
1 tsp boric acid

Yields: 2 toss

Time to make: About 5 minutes
Shelf life: 2 to 4 weeks
Storage: Use immediately.

Method:

Put jam in a shallow saucer, add boric acid powder and stir until combined.

How to use:

Place the mixture in areas with ant activity, or near mounds.

Note: Keep out of the reach of children and pets to avoid accidental ingestion.

ESSENTIAL ANT REPELLENT

This is an easy way to keep ants away from bench tops and other areas.

Ingredients:

Peppermint essential oil

Yields: As necessary.

Time to make: Instant
Shelf life: Indefinite
Storage: Keep in original bottle.

Method:

None!

How to use:

Sprinkle about 3 drops of peppermint essential oil along bench tops and baseboards.

Note: Keep out of the reach of children and pets to avoid accidental ingestion.

ANTS AWAY MIST

Keeps ants at bay the easy way!

Ingredients:

10 drops Pennyroyal essential oil
1 litre water
6 drops Peppermint, Spearmint or Citronella, alone or in any combination.

Yields: 1 Litre.

Time to make: About 5 minutes
Shelf life: Indefinite
Storage: Spray bottle.

Method:

In a spray bottle combine water and essential oils. Secure lid and shake vigorously.

How to use:

Spray liberally in areas with ant traffic.

Note: Pennyroyal is contraindicated in pregnancy. Consult your health care professional or herbalist before using. Keep out of the reach of children and pets to avoid accidental ingestion.

IN THE PANTRY ANT ELIMINATOR

This effective ant poison is simple to make and you are likely to already have most of the ingredients needed.

Ingredients:

6 tbsp peanut butter
1 tbsp brown sugar
½ tsp salt
1 tbsp boric acid

Yields: 8 toss.

Time to make: About 10 minutes
Shelf life: 4 to 6 weeks
Storage: Store extra in a glass jar with tightly fitting lid.

Method:

Put all ingredients in a glass jar and stir until well combined.

How to use:

Put peanut butter mixture in shallow dishes in areas frequented by ants.

Note: Keep out of the reach of children and pets to avoid accidental ingestion.

ARCTIC ANT MIST

Clean and refreshing ant preventative. This recipe is so simple and effective I am certain it will become a staple in your natural insecticide collection.

Ingredients:

2 cups white vinegar
2 tsp peppermint essential oil or your favourite mint oil

Yields: 2 cups.

Time to make: About 10 minutes
Shelf life: Indefinite
Storage: Spray bottle.

Method:

In a spray bottle combine vinegar and peppermint essential oil. Shake vigorously to mix.

How to use:

Mist vinegar solution along baseboards, under appliances, around door frames and widow seals, or anywhere you notice ant activity.

Note: Please test in an inconspicuous area before using. Keep out of the reach of children and pets to avoid accidental ingestion.

HONEY INSECTICIDE

This mixture is excellent for areas of high ant traffic.

Ingredients:

- **½ tsp honey**
- **½ tsp boric acid**
- **½ tsp aspartame (NutraSweet)**

Yields: 1 tsp.

Time to make: About 5 minutes
Shelf life: 2 to 3 weeks
Storage: Use immediately.

Method:

In a small glass dish or bottle combine honey, borax, and sweetener. Stir until well mixed.

How to use:

Place in areas of high ant traffic. This mixture should not kill worker ants immediately but instead will be carried back to eliminate the colony.

Note: Do not use this mixture outside. Keep out of the reach of children and pets to avoid accidental ingestion.

NATURAL CITRUS MIST

This 100% natural citrus mist works well for repelling ants and for cleaning areas where ants have been active.

Ingredients:

Peels from about 4 oranges or lemons
1 cup water

Yields: 1 cup.

Time to make: 3 to 4 hours
Shelf life: Indefinite
Storage: Spray bottle.

Method:

Place orange peels in a glass jar, bring water to boil. Pour water over orange peels and leave to rest for 3 to 4 hours. You can also put the mixture in a warm sunny window to increase its intensity. Remove orange peels and pour orange or lemon water into a spray bottle.

How to use:

Mist anywhere where ants are active including bench tops as well as around door and window frames. This mixture can also be sprayed directly on ants.

QUICK TIPS FOR ANT CONTROL

The following hints and tips will help you deal with ants at a moment's notice:

- When you see ants following each other in a line, they are actually following a chemical scent trail to a food source left by scout ants. You can deter ants by lightly misting the trail with a soap and water solution such as Pleasing Peppermint Soap Mist and then wiping off the excess. Follow this by placing a cinnamon stick across their path.
- Consider growing herbs such as mint or lemon balm near your kitchen window. You can also place potted herbs near known ant activity areas.
- Placing cotton balls which have been treated with a few drops of mint essential oil in areas where ants are active will deter them. Essential oils do evaporate, so it will be necessary to replace the cotton balls every day or two.
- Leave mint tea bags in areas with ant activity.
- Crush or finely chop dry mint leaves or cloves and sprinkle along ant paths.
- Seal any cracks around doors or windows.
- Make sure all windows have screens.
- Store pet food, sugar, grains and cereals in air-tight plastic containers or glass jars with seals.
- Keep ants out of your pet's water or food dishes by placing the bowl in a pie tin or other shallow dish filled with water and a few drops of soap. Caution: Make certain your pet does not drink the soapy water by covering it with mesh or placing it inside an empty bird cage.

- Cornmeal sprinkled around outside areas is a quick non-toxic solution to ant problems.
- Cut a fresh lemon (you can also use bottled lemon juice) in half and squeeze into and around the entrance of the ant mound. (Be careful not to be bitten!) You can leave the lemon rind near the entrance too.
- Do not throw away cucumber rinds. Ants do not like the bitterness of cucumbers and will avoid them when placed around the entrance of an ant mound or across an ant trail.
- Ants are repelled by the scent of pine. Try adding a drop of pine essential oil to your cleaning water. Wiping or mopping with this water will help deter ants.
- To prevent ants from invading picnic tables, set each leg of the table in a dish of water with a few drops of your favourite mint essential oil added.
- A solution of water and soap, such as Pleasing Peppermint Soap Mist, can be kept ready in a spray bottle. Whenever you see an ant, spray it with the solution. This also works for other bugs (use natural castile soap).
- Ants will avoid cinnamon. If ants are getting in under doors or windows, sprinkle a line of ground cinnamon, or a place a cinnamon stick, across their path to prevent them from entering.
- Ants will also avoid flour; however use caution when using flour, as it may attract other insects. To deter ants, sprinkle a line of flour along the backs of shelves or wherever you notice ant activity.

In addition to the natural ant barriers already mentioned, here are a few other natural ant deterrents which, when placed across ant trails, will prevent ants from passing:

- Coffee grounds
- Cayenne pepper
- Chalk
- Cucumber peels
- String soaked in lemon or orange essential oil
- Lemon juice

BEDBUGS

History and Habits

Until the 1940s bedbugs were man's consistent companions. The earliest natural history writings by Aristotle and Pliny mention them. In ancient times, the bedbug was thought to have medicinal value, and ground bedbugs were recommended to relieve everything from fainting spells to fevers. Thanks to our grandparents, most of us have never seen a bedbug but since 2003 their numbers have been on the rise. Hotels are especially prone to infestations; this is a problem as they can hide in clothing or suitcases and cause infestations at home upon return from a relaxing holiday.

Unfortunately, a bedbug infestation is almost difficult to eliminate and does require persistence due to the bug's lifecycle.

About the Bedbug

Bedbugs are members of the Cimicidae family and, like mosquitoes and fleas, feed on blood. When fully grown the average adult bedbug will measure 3mm. around and between 4 and 5 mm. long. They are oval in shape and very flat which means they can wedge into tiny crevices where they go undetected. Bedbugs are wingless and unlike mosquitoes or fleas, cannot fly or jump. They produce a pungent odour similar to stink-beetles, which once you have smelt, you will quickly recognise again.

Bedbugs do not go through a complete metamorphosis like butterflies or beetles. Instead they hatch from whitish coloured eggs (large enough to see without the aid of a microscope) looking like small adults, except they are a pale, translucent yellow. They turn a reddish brown and get bigger as they moult. Typically, a bedbug with a constant food source will moult every 8 days. However, if they cannot find a victim to feed on, they can wait 16 months between feeds. To become mature, a bedbug must feed five times on a host, and at least once more if they are to develop eggs. Female bedbugs can lay between 75 and 200 eggs in their lifetime.

Bedbugs are often found piled together in large groups, which is usually where they will deposit their eggs.

Bedbugs feed at night and spend the day hidden away in tiny cracks and crevices. They will often travel long distances to find a host and have been known to emerge in daylight to feed if they are hungry. Favourite hiding places for bedbugs include mattresses, crevices in wooden beds, between boards, under and along base boards, between walls and carpets, and any other location suited to their flat body shape.

Its uncanny ability to find peaceful sleepers and make a meal of them has made the bedbug an enormously unpopular pest. The piercing mouth parts are bent down when they are about to bite and slide past each other to puncture the skin of the victim. The bugs then begin to suck blood up through their tube-like mouth parts, and have been observed taking 70 draughts of blood a minute. A small amount of saliva is injected into the bite, which helps prevent the blood from coagulating. Bedbugs have a tendency to stay on clothing instead of clinging to the skin, which actually helps prevent wound contamination from excrement. The bedbug will draw blood for about ten to fifteen minutes before it becomes full. Like the tick, once it is full the body is no longer flat, but round and distended with blood. Once the bedbug has fed it returns to its hiding place.

Common bedbug usually only feeds on humans, those which commonly infest birds or bats are a different species to the bedbugs referred to here.

Don't let the Bedbugs Bite

Each person is affected differently by a bedbug 'bite' or puncture. Some people have no reaction at all, while more sensitive individuals may experience a stinging sensation, discomfort, and a hard, white swelling. Typically, swelling seldom occurs from bedbug bites, unless scratched or rubbed. Unfortunately, people who have little or no reaction to bites will probably have larger infestations before a problem is noticed.

If you are experiencing bite-like symptoms, make a search of the bed and surrounding areas. Bedbugs will be easily discovered if an infestation exists. If you do not find bedbugs, you should seek medical attention from a physician who stresses a holistic approach, to determine if you have a skin condition or allergy and if this is the case be sure to seek the assistance of a naturopathic practitioner.

So, how can you tell if you have been bitten by a bedbug? This is difficult to answer because bedbug bites are very similar to mosquito or other insect bites. If you suspect a bedbug infestation check the cracks and crevices in your mattress, around your bed, even a loose bit of wallpaper can house bedbugs. Look for shed skin, eggs (bedbug eggs are about 1 mm across and oval shaped) or the bugs themselves. Another sign of an infestation is an unpleasant odour similar to that of stink beetles.

If you feel a bite or stinging sensation during the night, turn on a light and inspect the affected area. Bedbugs feed for long periods, so if you look quickly you should see them on your body, clothing or bed.

Common Myths about Bedbugs

Myth: Bedbugs bury themselves under the skin.

Fact: Bedbugs do not bury their heads or mouth parts under the skin. Bedbugs bite and then leave the host; they do not remain attached and do not require any special methods of removal. While they may resemble large ticks, think of them more as a wingless mosquito.

Myth: Bedbugs can occur spontaneously in filth.

Fact: This is certainly not true, unlike cockroaches, which feed on garbage and are attracted by dirt or unsanitary conditions; bedbugs feed on blood and are not attracted by dirt. Even the well-kept household is susceptible to bedbug infestations but keeping the home clean and well vacuumed will aid considerably in prevention and early detection.

Myth: Bedbugs travel long distances between houses to feed.

Fact: While this can be true, especially if a house has been deserted and the bugs need to find a new food source, it is not typical. Bedbugs like to make their nests near a food source, such as in the mattress.

BASIC BEDBUG CONTROL

Bedbugs cannot be treated with a bait program, because they feed on human blood. The most dependable method is inspection and subsequent treatment of infested areas and furniture.

A step-by-step investigation

The best place to begin is with a thorough inspection of the sleeping area. This includes the mattress, box springs, bedding, bed, bedside table, floor, walls, etc. Bedbugs can and do literally hide anywhere. They can hide behind base boards, under bits of loose wallpaper, or even small cracks between the wall and ceiling.

The Bed

The bed needs special attention as it is one of the mostly likely places where bedbugs will hide.

Disassembling and Cleaning the Bed

Start your search with the bed, you probably won't find anything on the top mattress or bedding, but you should inspect these none the less, as you may discover bedbugs hiding in the seams or under buttons. Carefully look at both the top and underside of the mattress. Brush both sides along the seams, around buttons, and in crevices with a stiff bristled brush, set the top mattress aside.

Once you have thoroughly investigated the mattress, sheets and bedding, wash the sheets and any washable bedding, such as a mattress cover. If you have any bedding which is not washable hang it out to air. Airing on very hot days in the sun or when it is freezing will likely kill any bedbugs still on the material.

Hint: You should frequently remove bedding and launder it in the hottest water setting allowed by the manufacturer. Bedbugs are susceptible to hot temperatures and not likely to survive when washed in hot water (at least 36-40 degrees Celsius and above is ideal).

Next, inspect the mattress base or box springs. If you have an infestation of bedbugs in your bed, this is where they will most likely be living, so look carefully. Inspect for fabric rips, a tear in the box spring fabric would open a door for bedbugs to find their way into the wooden frame. If you find a tear, make sure you thoroughly inspect the inside of the box spring as well. Remember that bedbugs have a pungent odour, so keep this in mind when searching.

If you find bedbugs in your mattress or box springs, you will need to follow a treatment program, which I will describe later, to remove them.

The Bed Frame

While you have your mattress and box springs disassembled it is a good idea to turn your attention to your bed frame. Start with a detailed vacuuming of your bed frame, always paying close attention to cracks and crevices. Finish by spraying with an all-purpose bedbug mist.

Reassembling and Finishing touches for the Bed

Before reassembling your bed, thoroughly vacuum the mattress and box spring. If you find bedbugs, vacuum them up as well. You can kill any bedbugs you vacuumed up by placing the vacuum bag in a plastic storage bag, and freezing it before throwing it away. Bedbugs that have fed are sensitive to freezing temperatures, however newly hatched bedbugs require longer exposure. To be sure all bedbug life stages are killed, write the date on the bag and keep it frozen for about 20 days. Unfortunately, bedbug eggs are very difficult to dislodge and are often well hidden. Even a thorough vacuuming may not remove them. This is why you will need to follow an extended treatment program to remove any further eggs which hatch or any missed females. The eggs area problem because there are no products that can kill them. Even chemical pesticides cannot, regardless of any claims made on the packaging. This is another good reason never to consider using chemical pesticides.

After you have vacuumed, mist the mattress and box spring with citrus bedbug mist (see recipes). Pay special attention to areas where you discovered any bedbugs. Mist even if you found no bedbugs to help prevent any bugs from making your mattress their home in the future. Be sure to give special attention to any cracks, seams, or tears, especially if you have discovered bedbugs during your inspection. Your mattress and box spring are now ready to be reassembled with fresh sheets and bedding.

Around the Bed

Bedbugs are not confined to living in beds, they can live anywhere in the room, near the bed, in the carpet, between wood boards if you have wood floors, in electrical outlets, in night stands, drawers, closets, etc., they will even climb to the ceiling to hide, so be especially careful if you have crown moulding.

The next place to inspect is the area around your bed; this includes the floor, wall areas near your bed, bedside table, etc.

Crown Moulding

Next, inspect mouldings. Begin at the top, if you have crown moulding make a careful inspection and then vacuum where the moulding and the wall meet. Finish by puffing natural pyrethrum powder or natural diatomaceous earth into any cracks.

Baseboards

Check along baseboards, taking special note of any spaces or cracks. Vacuum the edges where the baseboard meets the floor and the wall. Then puff natural pyrethrum powder or natural diatomaceous earth into any cracks and where baseboard meets the floor.

Switch plates, electrical outlets

Remove switch plates and puff in pyrethrum powder or natural diatomaceous earth. Be very careful not to touch the wires as they could cause electrical shock.

Wooden Furniture

Use the following steps to inspect and treat any wooden furniture such as dressing tables, cabinets, dressers, desks etc for bedbugs. If the furniture has drawers, begin by taking all the drawers out. Take any contents out of the drawers or shelves and look carefully for any bugs or signs of activity. Next, carefully inspect any joints and cracks. Vacuum all cracks, joints, and crevices, if there is no bug traffic, mist lightly with a bedbug mist and allow to dry. If you find bugs, omit misting and finish by puffing pyrethrum powder or natural diatomaceous earth into the joints and crevices and areas of activity. Replace all the drawers. Wash any clothing that has been stored in the furniture in the hottest water allowed by the manufacturer.

Chairs and Upholstered Furniture

Inspect and treat wooden chairs in the same manner as above. For upholstered furniture such as chairs or couches, begin by removing and vacuuming the cushions then vacuum the chair thoroughly. As always, pay particular attention to seams, cracks, gaps, and tears. You can finish by sprinkling the chair or couch with Earth powder (see recipes), and vacuuming up after the powder has sat for about 3 hours.

Wooden Floors

Wooden floors are the easiest to treat. People who have wooden floors are much less likely to have problems with the floor area around their bed, especially if the spaces between boards are sealed, or tight. If you are lucky enough to have wood floors, just keep your floors clean and check any cracks for bugs. If you do find a bug you can puff a bit of natural pyrethrum powder or natural diatomaceous earth into the crack and then vacuum using the hose attachment of your vacuum cleaner.

Carpeted Floors

Carpeted floors are more of a challenge, because the carpet fibres provide bedbugs with very good hiding places. Replacing your carpeted floor with a wood one is probably the best way to reduce bedbugs around your bed. However, carpets can be treated relatively easily.

To treat carpets: First, thoroughly vacuum and then apply a floor powder, such as Earth Power (see recipes), over the carpet. Let the powder sit for about 3 hours and then vacuum.

Extended Control Program

If you have discovered bedbugs you will have to repeat all the above inspection and cleaning steps two more times during the month, to make sure you have eliminated all of the bugs and their eggs. Any eggs that were missed should hatch out in about 7 to 10 days. Sadly, waiting for them to hatch is the only way to make sure you get all the bedbugs. Follow the steps again for inspecting, treating, and vacuuming your room. Then repeat the same steps again in approximately another 7 to 10 days. It is a lot of work, but it is worth the effort as bedbugs are a very unpleasant pest. Once you have completed these steps another two times your extended treatment is complete. However, it is still possible for another outbreak to occur. If this should happen, or if you have an infestation, you feel uncomfortable handling; never postpone contacting a professional for advice and treatment options.

Once you have completed the extended treatment you will need to carefully observe the results. You should notice a significant decline in bites and bug activity with no activity in about one to three weeks. It is not unusual to have another infestation after you have followed the treatment guidelines. Persistence is the most important element in controlling bedbugs as they are very difficult to remove. So if the first treatment doesn't completely kill them, just try it again! You may need to repeat the treatment steps a few times before you are completely free of bedbugs. Monitor your rooms for a few months and if you don't notice any activity you have probably successfully conquered the infestation.

BEDBUG CONTROL RECIPES

CITRUS BEDBUG MIST

This mixture can be misted on to mattresses, box springs, walls, and other areas bedbugs are likely to hide.

Ingredients:

- **2 cups water**
- **2 tsp Orange essential oil**

Yields: 2 cups

Time to make: About 5 minutes
Shelf life: Indefinite
Storage: Spray bottle.

Method:

In a spray bottle combine water and orange essential oil. Shake well to mix.

How to use:

Mist solution anywhere bedbugs are a problem or mist on potential problem areas as a preventative.

Note: Remember to test the solution in an inconspicuous area before using.

EARTH POWDER

You can sprinkle this powder on carpets or furniture to kill bedbugs. Because bedbugs squeeze into cracks, the powder often doesn't get to them. It works best in areas with exposed gatherings of bedbugs.

Ingredients:

- **1 cup bicarbonate soda**
- **1 cup cornstarch**
- **2 cups diatomaceous earth (natural, not pool grade)**

Yields: 4 cups

Time to make: About 5 minutes
Shelf life: Indefinite
Storage: Screw top glass jar.

Method:

In a glass jar combine bicarbonate soda, cornstarch, and natural diatomaceous earth. Secure lid tightly and shake to mix.

How to use:

Sprinkle onto upholstered furniture, carpets, or over areas of exposed bedbugs. Allow to sit for about 3 hours, and then vacuum.

DIATOMACEOUS EARTH

This works especially well for killing bedbugs if it comes in direct contact with them. One of the best things about diatomaceous earth is that is can remain effective for up to a year!

Ingredients:

Diatomaceous earth

Yields: As required.

Time to make: None!
Shelf life: Indefinite
Storage: Container with tightly fitting lid.

Method:

None required.

How to use:

Shake a fine dust of diatomaceous earth wherever you are having difficulties with bedbugs. In cracks, crevices, or seams are good places to apply diatomaceous earth.

Note: Be certain you purchase natural diatomaceous earth and not pool grade. In addition, diatomaceous earth is a fine powder and may enter electronic devices such as computers, stereos, or gaming consoles, and cause damage. Cover nearby electronic equipment before dusting with diatomaceous earth.

HINTS AND TIPS FOR BEDBUG CONTROL

- Bedbugs are becoming an increasing issue in hotels and motels. Pack a disposable mattress cover when travelling and use it to cover the hotel mattress. Be sure to throw it away at the hotel and don't repack it.
- Bedbugs are sensitive to heat, if you heat an infested area to 40 degrees they should be killed in large numbers. During hot weather, consider vacating your home for a couple of days. Turn off all air-conditioning and close all the windows and doors to your home. This will help keep heat in and the temperature high.
- Frequent vacuuming will help eliminate bedbugs before they have a chance to take hold if you have an infestation; this also reduces the number of adult bedbugs.
- Stop the infestation before it starts! Don't bring infested items into your home. This means you should never buy used furniture, mattresses, or bedding without first thoroughly examining it. One way to this is to check for bedbug excrement, which appears as tiny reddish-black flecks.
- Caulk all cracks leading into your home and make sure all windows have screens.
- Wash clothing and bedding frequently in hot water (40 degrees Celsius).
- Place vacuum cleaner bags in a dated plastic storage bag and freeze for about 20 days. This will kill any bedbugs in the bag and help prevent future infestations. Freezing eliminates the chance of bedbugs escaping from the garbage and reinfesting yours or a neighbour's home and it will kill their eggs.
- A stiff bristled brush used along the seams of furniture and mattresses will help dislodge bedbug eggs.

- Throwing out your infested mattresses can help reduce an infestation; however you should only throw out your mattress if the bedbugs have been removed from the rest of the room.
- Make a barrier around your bed by wrapping double-sided tape around the bed's legs.
- Stop bedbugs from crawling up bed legs by setting bed legs in shallow dishes filled with soapy water (if possible). Since bedbugs can't fly or jump they have to crawl up the bed legs to feed, and the soapy water barrier will prevent them from climbing.
- Bedbugs are most active at night, before going to bed check your sheets, pillows, and bedding for any active bedbugs.

CARPET BEETLES

History

In the United States, this bug is also known as the buffalo beetle or buffalo moth, possibly because it was discovered in Buffalo, New York. Another historic suggestion was that the name originated from the furry lint coating the larva acquires as it feeds on wool, making it slightly resemble a buffalo.

About the Carpet Beetle

The carpet beetle larva has a voracious appetite and loves wool fibres however, it will also eat silk, fur, leather, cotton, feathers, books, and other natural fibres. While clothing made from synthetic fabrics is not vulnerable, larvae may be found on synthetic natural blends. The larvae do not stick to fabrics and may be easily brushed off.

A carpet beetle may be mistaken for a lady beetle at first glance. It is a little less than 6 mm long, elliptical in shape, and covered in tiny scales, which make it look marbled black, yellow or white, depending on the species. The head is small and difficult to distinguish from the rest of the body except by two short jointed antennae. Each leg ends with a tiny hook. If disturbed, the carpet beetle will play dead. Carpet beetles can be seen during the day flying around. Once they have shed their skin, typically in summer, they try to leave the house to look for the pollen of their favourite flowers. For this reason they are often found on window sills and draperies.

Where larvae have been eating carpets, you will sometimes see irregular holes or long slits. If a garment is folded, you will often find small holes eaten at each fold of the fabric.

The less disturbed and darker areas of wool carpets make excellent living, breeding, and feeding places for carpet beetles. Frequent vacuuming can keep them from taking hold in wool carpets.

Controlling Carpet Beetles

Carpet beetles are the worst enemies of wool, and once they have established themselves in a carpet or closet it will require a very persistent effort to eradicate them. Worse still, a carpet beetle infestation may go unnoticed until they have caused significant damage. It may be necessary to treat your home three or four times before the infestation is eliminated.

Treatment Program for Carpet Beetles

Before following the treatment steps, vacuum all the carpets in your home thoroughly. This reduces the number of beetles and larvae in the carpet and brings any that were missed closer to the surface, making them easier to remove.

Treating Carpets

Begin by treating all of the carpets in your home (this includes rugs, runners and mats) by liberally sprinkling carpet powder (see recipe in carpet beetle control recipes) over rugs.

It is important to treat all of your natural fibre carpets, even if they do not show signs of an infestation, because larvae can hide deep in the fibres and go undetected until they turn into adults.

If the rug is small, you can take it outside and beat it, which should dislodge any larvae or beetles living in the fibres. Pull back carpet edges and puff diatomaceous earth at least 1 foot underneath.

Regular vacuuming will help prevent carpet beetles from infesting carpets and reduce any current infestations.

Treating Closets

Inspect all clothing and then shake vigorously over a plastic garbage bag. Follow by thoroughly vacuuming the floor of the closet. Another way to remove larvae from clothes is to toss the infected item into the dryer. If a garment is badly infested you may have to throw it away.

Treat any closets where you have noticed activity by misting the floor with a soapy water mixture, such as Pleasing Peppermint Soap Mist (see general pest control recipes) and then puffing diatomaceous earth around the baseboards and any areas where you have noticed carpet bug activity.

Treating Upholstered Furniture

Begin by removing any cushions and thoroughly vacuuming them, and the rest of the item. Finish by sprinkling the furniture with a powder such as Carpet Beetle Control Powder (see carpet beetle control recipes), then vacuum up the powder after it has sat for about 3 hours.

Continuing Treatments

Carpet beetle eggs and pupae are difficult to kill, so it may be necessary to repeat the treatment three or four times over the next month, before all eggs, larvae, pupae, and beetles have been eliminated.

CARPET BEETLE CONTROL RECIPES

CARPET BEETLE CONTROL POWDER

You can sprinkle this powder on carpets or furniture to kill carpet beetles.

Ingredients:

- **1 cup bicarbonate soda**
- **1 cup cornstarch**
- **2 cups diatomaceous earth (natural, not pool grade)**

Yields: 4 cups

Time to make: About 5 minutes
Shelf life: Indefinite
Storage: Screw top glass jar

Method:

In a glass jar combine bicarbonate soda, cornstarch, and natural diatomaceous earth. Secure the lid tightly and shake to mix.

How to use:

Sprinkle liberally onto carpets or furniture; allow sitting for about three hours, and then vacuum.

BEETLE ELIMINATOR MIST

Ingredients:

- **1 litre water**
- **½ cup Palm oil**
- **3 drops Citrus essential oil (your choice)**
- **3 drops Neem essential oil**

Yields: 4 cups

Time to make: About 5 minutes
Shelf life: Indefinite
Storage: Screw top glass jar/spray bottle

Method:

Combine palm oil, citrus essential oil, neem essential oil, and water in a glass jar. Secure lid and shake vigorously to blend. Pour into a spray bottle for easy misting.

How to use:

Spray mixture on any visible beetles.

Note: Over 200 insect species are known to be negatively affected by Neem essential oil. So you can effectively try this recipe on lots of other bugs!

Tips and Hints for Controlling Carpet Beetles

- If you have a closet or attic where woollen or silk items are stored, try laying out a piece of red wool flannel. Every few days pick up the piece of flannel and after washing it, soak it in boiling water for a few minutes, let it dry and then replace it. The flannel will attract the bugs, and help keep your valuable fabrics safe.
- Store susceptible fabrics, such as winter sweaters, in plastic bags to protect them from carpet beetles.
- Apply some Neem essential oil onto some cedar chips and place them on closet shelves, hang in sachets on clothes hangers or place them in corners.

COCKROACHES

History

Cockroaches are found in all climates and countries. They belong to one of the oldest insect families in the world, even the dinosaurs had to co-exist with them. And despite all our advances, we still share our homes with this persistent pest.

About the Cockroach

There are four main species of cockroach and all are identifiable by colour and size. Oriental cockroaches are shiny and very dark brown, almost black. The females are almost wingless but the males have thin wings and are short and plump. American and Australian cockroaches are both lighter in colour, ranging from chocolate brown to a very light caramel colour, but Australian cockroaches have two dark marks on the centre of their bodies. American cockroaches have darker bands or spots near their heads, at the base of their wings and on their backs. German cockroaches are a light caramel colour and much smaller than the other species.

For all cockroaches, eyes are of little or no significance, their legs are well suited to jumping and scurrying, and their antennae are very long. They are very active and will quickly run away if a light is turned on or they sense a disturbance. A cockroach is more likely to scurry, than fly, even though it has wings. Their abdomens are made of overlapping segments, which can be stretched out to make their bodies flat and thin so they can fit through very small cracks.

Cockroach eggs are laid in brownish coloured, enclosed, oval shaped capsules, called oothecae, which are banded with darker coloured rings. If you discover any of these egg cases, they should be burned.

All life stages of cockroaches may be considered nuisances as they are active and invade homes. Larvae are white when they hatch, and quickly turn yellow and blackish brown. Both the larva and pupa look like small wingless adults. Young roaches are often seen with an adult, which has led to speculation that the adults may nurture the young for a short time after hatching.

Cockroaches love to feed on any rotting organic matter, such as food scraps. However, they will also eat dry or fresh food and pet waste. They will even chew on leather, fabric, and paper, and the damage they cause is often irreversible. Poorly rinsed dishcloths and sponges are a favourite place for cockroaches to feed so it is important to keep dishcloths well rinsed and clean. Cockroaches contaminate food. They have been linked to cases of salmonella poisoning, bubonic plague, E. coli, pneumonia, and many other devastating diseases. The same cockroach may have come up from the sewer, eaten from the garbage or cat litter box, and then made its way to your silverware or dishes. Any dish or utensil on which you have noticed a cockroach must be thoroughly washed immediately.

The amount of food cockroaches eat, while not inconsiderable, would likely go undetected, if they didn't frequently ruin the food they come in contact with and leave a disgusting odour wherever they have been feeding. This smell comes from their mouths, excrement, and scent glands. Any areas where cockroaches have been living have to be washed and disinfected, to remove the scent (this can be done using a soapy water solution and a mist contain tea tree oil) and any clothes that have been in contact with roaches will have to be washed to remove the smell. Cooking utensils stored in drawers frequented by cockroaches must also be thoroughly cleaned, as they can pass the odour on to food and drinks, as either a taste or smell.

The odour is so distinctive in fact, that a sensitive individual, familiar with the smell, may immediately detect an infestation of cockroaches in a home or public building, even if they can't be seen.

CONTROLLING COCKROACHES

Cockroaches love to make their homes in warm, damp, dark areas, such as walls with hot water pipes, or under kitchen counters. Once they have become established it will require persistent treatment and ideal cleanliness to eliminate the infestation. Fortunately, bait programs are usually very effective, especially when combined with eliminating the cockroaches' food sources, keeping the home perfectly clean, and repairing any leaky pipes.

Tips for Cockroach Prevention

- Make sure all kitchen scraps are thrown away and counters are wiped clean. It is also a good idea to empty kitchen bins at the end of each day.
- Make sure any compost containers are tightly sealed.
- Prevent insects from entering your home by eliminating cracks or holes leading to the outside. Check around your home for any places cockroaches can enter, such as around the foundations, through electrical boxes, and badly sealed door or window frames. Caulk any gaps that you can, and sprinkle a natural cockroach deterrent such as cayenne pepper, mint, diatomaceous earth, or boric acid around the area.
- Place roach deterrents such as mint, cayenne pepper, or cotton balls soaked in citrus or tea tree oil behind kitchen appliances, furniture, in drawers, shelves, closets, and in the pantry.
- Do not be discouraged if one repellent doesn't work, you may need to try a few before finding the one that works best. Remember cockroaches are renowned for their persistence.

THREE STEP ROACH CONTROL PROGRAM

This comprehensive program consists of an inside treatment, an outdoor treatment and a garden treatment.

Step 1: Inside Treatment

The ideal treatment inside your home is to use baits. Baiting works especially well because cockroaches live as a community. They eat poisoned food from a bait, return to the nest and die. Other cockroaches eat their remains and consequently also die. This circle continues and kills cockroaches where they breed.

It is important to note that baiting works best when it is the cockroach's only food source. Make sure all garbage is removed, bench-tops and floors are free of crumbs, and all food packages are sealed tightly.

Once you choose a bait recipe, be sure to place as many baits around your home as needed. You may need to experiment to find the recipe that appeals the most to your cockroaches. Good locations for baits include behind furniture, behind kitchen appliances, under sinks, and anywhere you have noticed cockroach activity.

Check the baits about once a week and refill if necessary. It is a good idea to keep a few baits around your home even if you are not experiencing an infestation. In that case, the baits only need to be checked every four to six weeks to make sure they haven't evaporated.

Step 2: Outdoor Treatment

The outdoor treatment focuses on treating the perimeter of your home. For best results you should treat the outside of your home on a monthly or bimonthly basis, or after heavy rains, to prevent new infestations.

An outdoor barrier stops pests coming in from garbage areas and other breeding places. To create an outdoor barrier, draw a line around the foundation of your home using cayenne pepper or any other cockroach barrier such as boric acid. Follow by treating any cracks in the foundation with boric acid or insect power.

Step 3: Garden Treatment

Finally, you need to tackle outdoor sources of cockroaches by careful baiting.

Begin by looking for cockroach breeding grounds, such as under damp logs, or near rubbish bins. If you locate a group of cockroaches place a bait near it, or sprinkle a mixture of boric acid and sugar near the area. The cockroach solution recipes have more mound treatment ideas.

Place baits behind shrubs and bushes, near garbage bins, and in any areas of known pest activity. Pests are often attracted to moist, damp, or dark areas, so consider placing baits in these areas.

You may need to try several recipes or control solutions before you find the one that appeals to your cockroaches. As with any pest control program, you may need to adjust recipe concentration with varying individual results.

Reminder: Be sure to keep baits away from areas accessible to pets and children—indoors or out.

COCKROACH CONTROL RECIPES

APPLE ROACH TRAP

This trap uses beer to lure and drown roaches.

Ingredients:

Beer
1 piece of banana or apple
Tape
Nonpetroleum jelly
Glass jar

Yields: 1 ½ cups

Time to make: About 5 minutes
Shelf life: Approximately 1 month
Storage: Glass jar

Method:

In a glass jar place the piece of banana and fill jar about half way full with beer. Rub non-petroleum jelly around the inside rim of the jar (this makes it too slippery for the roaches to climb out). Wrap the jar in masking tape, or tape newspaper around the outside, so that the cockroaches can climb up the outside.

How to use:

Place jar in an area where cockroaches are likely to discover it, such as under the kitchen sink. The roaches will be lured by the banana or apple, climb up the jar, fall in and drown.

COCKROACH HERBAL MIST

This mist uses herbs and essential oils that repel cockroaches.

Ingredients:

2 cups water

2 teaspoons of your favourite herb or essential oil from the list below:

- **Tea tree essential oil**
- **Peppermint essential oil**
- **Crushed hot pepper**
- **Garlic**
- **Bay leaves**
- **Peppermint (herb)**

Yields: 2 cups

Time to make: About 5 minutes
Shelf life: Indefinite
Storage: Spray bottle

Method:

Combine the herb or essential oil of your choice and the water in a spray bottle. Secure the lid and shake vigorously to blend.

How to use:

Mist in areas where there has been cockroach activity; wipe dry using a paper towel.

COCKROACH MOTEL

This recipe has long been considered one of the most effective for cockroach control. It makes several baits.

Ingredients:

1 cup sugar
1 cup boric acid

Yields: 2 cups

Time to make: About 5 minutes
Shelf life: Indefinite
Storage: Glass jar with tightly fitting lid

Method:

In a glass jar, combine sugar and boric acid. Secure lid and shake vigorously to blend.

How to use:

Shake about 1 tablespoon of the mixture into a shallow dish or cockroach box (see directions for making one below). Place in areas frequented by cockroaches such as behind kitchen appliances and under sinks.

Variations:

Mix 1 teaspoon boric acid with about 1 teaspoon of chocolate shavings.

Note: Keep out of the reach of children and pets to avoid accidental ingestion.

HOW TO MAKE A COCKROACH BOX

To make a cockroach box: Cut a small hole at each end of a match box (or other small box). Cockroaches enter through the holes in the box and eat the bait.

COCKROACH IN PARIS

This is another tried and true recipe for cockroach control that has been used for over a century.

Ingredients:

½ cup plaster of Paris
½ cup flour or cornstarch or sugar

Yields: 1 cup

Time to make: About 5 minutes
Shelf life: Indefinite
Storage: Glass jar with tightly fitting lid

Method:

In a glass jar combine plaster of Paris and flour. Secure lid and shake well. Shake mixture into a shallow dish.

How to use:

Place a shallow dish filled with the mixture in areas where cockroaches are likely to discover it, place another dish filled with water nearby. Cockroaches eat the plaster of Paris and flour mixture, which makes them thirsty, so they want to drink water. When they drink the water, it mixes with the plaster, which becomes hardened in their intestines, resulting in death.

COCKROACH BE-GONE

Another extremely effective recipe against these unwelcomed visitors.

Ingredients:

- **1 tablespoon insect powder (natural pyrethrum powder)**
- **2 tablespoons natural diatomaceous earth (no pool grade)**

Yields: 3 tablespoons

Time to make: About 5 minutes
Shelf life: Indefinite
Storage: Glass jar with tightly fitting lid

Method:

In a glass jar combine insect powder and natural diatomaceous earth. Secure lid and shake well.

How to use:

Dust areas where cockroaches have been seen.

CAJUN COCKROACH MIST

Add a little spice to your cockroach treatments. Keep a bottle of this handy to mist any wandering cockroach you find.

Ingredients:

- **4 cups water**
- **1 tablespoon Tabasco sauce**
- **1 tablespoon liquid Castile soap**

Yields: 4 cups

Time to make: About 5 minutes
Shelf life: 1 to 2 months
Storage: Spray bottle

Method:

In a spray bottle combine water, hot sauce, and Castile soap. Secure lid and shake well to blend.

How to use:

Mist into areas of known cockroach activity, the spices act as a deterrent, or spray directly on any cockroach you see to kill them.

RELIABLE COCKROACH DESTROYER

Nothing works better than this old standby.

Ingredients:

Boric Acid

Yields: As required.

Time to make: None!
Shelf life: Indefinite
Storage: Original container

How to use:

Lightly dust boric acid behind kitchen appliances, along baseboards, under sinks, and in cracks. Cockroaches love high places so sprinkle on the top of kitchen cupboards, the refrigerator, and other high locations.

Note: Keep out of the reach of children and pets to avoid accidental ingestion.

Tips and Hints for Controlling Cockroaches

- Place cotton balls treated with six to eight drops of essential oil of citronella at the back of cabinets or in drawers. To increase the repelling properties of the cotton balls add a drop of lemongrass or peppermint essential oil. Another alternative is to use a few drops of mint essential oil on a cotton ball.
- Cockroaches are repelled by cucumbers. Place slices or small chunks of cucumber around areas where you want to prevent them from entering.
- Cockroaches have a natural aversion to catnip. Grow a pot of catnip in your kitchen (or garden) or place catnip sachets in areas with cockroach traffic. Alternatively, make a catnip mist by simmering fresh catnip in a pot with water. Once the catnip and water have steeped for a few minutes, pour the water into a spray bottle and use it to mist areas where there is cockroach activity.
- Keep a spray bottle filled with a soapy water solution for misting any wandering cockroaches. As an on-the-spot spray, this is just as effective as poisonous commercial products.
- Check the defrost pan under your refrigerator about once a month. Before putting it back, wash and dry it and add a few drops of pine essential oil.
- Make sure you check for leaky pipes and taps. If you find any you should repair them quickly because cockroaches are drawn to moisture.
- Make sure all windows are screened.
- Covering kitchen drains with fine mesh screens will not only stop cockroaches from entering through drains but also from exiting. Make sure the mesh is very fine; cockroaches are flat and can slip through slotted strainers.

- Cockroaches love to feed on poorly rinsed dishcloths and sponges so always rinse thoroughly. For added protection, you can place them in a sealable plastic bag with a drop of pine essential oil.
- Several essential oils such as clove, mint, neem tree seed, thyme, rosemary, and limonene (which is extracted from citrus peels), have all proven toxic to cockroaches. Some of these super oils are also available in dust form. Placing a cloth, soaked in one of these oils or making a special sachet with them and placing it in the pantry will deter cockroaches from entering. Keep replacing the oils as they evaporate with time.
- Cockroaches can easily slip into food stored in its original cardboard box or wrapper. Try to store foods like cereals, flour, sugar, etc., in airtight containers or sealed plastic bags.
- Any open sources of water in your home can draw cockroaches, be sure to check the water trays under potted plants and make sure there is no standing water.
- Because cockroaches thrive in warm moist areas, you may want to consider a dehumidifier.
- It is a good idea to keep a few sticky traps around to monitor cockroach activity. Check the traps frequently and place baits near high activity areas.
- Do not leave any dirty dishes over night.
- Frequent vacuuming will also help reduce cockroaches.

DUST MITES

About the Dust Mite

Dust mites live in air, furniture, carpets, clothes, and anywhere dust collects. While they may be invisible to the naked eye, these microscopic mites are actually one of the leading causes of allergic reactions. Only plant pollen ranks higher as an allergen. Dust mites shed their skins as they grow and those skins, the dust mites themselves, and their faeces, are so light they actually float in the air we breathe.

When viewed under a microscope, dust mites resemble ticks. They are from the arachnid insect family, which also includes spiders and scorpions.

Dust mites don't bite people, or ruin food but sensitive individuals are often adversely affected by breathing them in. Allergic reactions can become especially bad with larger populations of dust mites, usually due to a build-up of dust. Increasing the frequency of vacuuming and dusting will quickly reduce the number of dust mites in your home, and often bring fast relief for those who suffer from dust mite related allergies.

It is important to remember that allergic reactions can be caused by a wide range of allergens. If you have persistent allergy symptoms you should check with your health care professional to find out their cause. They may even be cause by a chemical sensitivity, which makes this another good reason to implement a chemical free lifestyle.

Dust mites eat the dander and flaked skin shed by people and animals. They will gather and breed in areas where these organic particles accumulate, such as mattresses, cushions, etc. Dust mites often collect in pillows, they simply live and breed inside the case without ever leaving. Individually they are extremely light but over time, as they breed and shed their skins, pillows can increase by as much as ten percent in weight. Mattresses are another place requiring special attention as they can harbour millions of dust mites.

During their larval and post-larval stages, dust mites have eight legs like spiders but they lose them by the time they are adults. Once dust mites hatch, they begin eating. They do not actually have stomachs and digest their food outside their bodies, using an enzyme that breaks down the food. The same piece of food is eaten several times before it is completely digested but it is never actually ingested. The left-over digested food is referred to by scientists as dust mite faeces. Bits of partially digested and fully digested food also contribute to allergies.

Dust mites have a life span of about 30 days for males and 90 days for females. During the last month of her life, the female will lay approximately 100 eggs. Dust mites are very sensitive to humidity levels, if the humidity becomes too low dust mites dry out and die because they do not drink water, they draw moisture from the air.

Dust mites are sensitive to direct sunlight, so opening the windows to let in fresh air and sunlight will contribute greatly to reducing dust mite populations.

Controlling Dust Mites

Because dust mites are so small, it is impossible to completely eradicate them. So, while the thought of these tiny creatures crawling about in pillows and mattress may be unpleasant, they are unfortunately here to stay. They have co-existed invisibly with us for thousands of years. It is only with the invention of the microscope 300 years ago that we discovered them. The best control program for dust mites is simply regular cleaning and vacuuming, which will greatly reduce their numbers.

Treatment Program

Look around your home for the places dust mites are most likely to collect, paying particular attention to any accumulated dust. A thorough natural cleaning program is the best way to reduce dust mite populations and protect the health of your family.

Vacuum

Regular vacuuming is the most important step because dust mites live and thrive in dust and areas where shed skin flakes collect.

The Bed

Since we typically spend about eight hours each night in our beds, they become one the main places for skin particles to collect. This is why bedding, mattresses, and pillows need our special attention.

Bedding should be frequently removed and washed; dust mites will be rinsed out of the bedding during the wash cycles. A thorough vacuuming of the mattress will remove flaked skin, dust mites, and their faeces and a thorough vacuuming of low traffic areas, such as under the bed, furniture, and anywhere dust collects, will also greatly reduce the dust mite populations.

Steam Cleaning

Steam cleaning mattresses, cushions, pillows, pet beds, carpets, plush toys, curtains, etc, is an excellent way to eliminate dust mites. The high temperature kills the mites and it is completely non-toxic. The moisture added to items can be an issue as it may foster future dust mite populations, so dry items in the sun, or with a hair dryer.

Steam clean your most vulnerable items as often as needed. If you find it successful (and you will know by decreased allergy signs and symptoms), it may be worth investing in a home steam cleaner.

DUST MITE CONTROL RECIPES

DUST MITE MIST

Use this mist on furniture to reduce dust mites.

Ingredients:

- **4 cups water**
- **½ cup boric acid**
- **1 drop pine essential oil**

Yields: 4 cups

Time to make: About 5 minutes
Shelf life: Indefinite
Storage: Glass jar, spray bottle

Method:

Combine water, boric acid, and pine essential oil in a glass jar. Shake vigorously until the boric acid has dissolved. Pour half the mixture into a spray bottle.

How to use:

Mist onto areas where dust mites are likely to live. For best results, apply every two months.

EUCALYPTUS WASH BOOSTER

Reduce even more dust mites with your next wash by using this natural dust mite eliminator.

Ingredients:

1 drop neem essential oil
6 drops eucalyptus essential oil

Yields: 4 cups

Time to make: None!
Shelf life: Indefinite
Storage: Original bottles

How to use:

Add the drops directly to the rinse cycle of your washing machine.

UNDER THE BED MIST

Mist this formula under any furniture, especially beds, to reduce dust mite populations.

Ingredients:

- **3 drops eucalyptus essential oil**
- **2 cups white vinegar**

Yields: 2 cups

Time to make: About 5 minutes
Shelf life: Indefinite
Storage: Spray bottle

Method:

In a spray bottle combine vinegar and eucalyptus essential oil. Secure lid and shake vigorously to blend.

How to use:

Mist under beds or other low traffic areas where dust is likely to accumulate.

Tips and Hints for Controlling Dust Mites

- If you live in an area with high humidity, in which dust mites thrive, it may be worth purchasing a dehumidifier. Running your air conditioner in summer and heater in winter will also help keep the humidity low and dust mite populations to a minimum.
- Wash or steam clean curtains and other fabric window coverings regularly to keep them dust mite free.
- Hardwood and tile floors are easier to keep clean than wall-to-wall carpet—although it is in your and your partner's best interest not to be exposed to any sort of renovations for at least 120 days prior to and throughout conception attempts and pregnancy. However if you have an extremely severe, known reaction to dust mites you may consider replacing carpeting with one of these alternatives, as it can greatly reduce the area where dust mites can live.
- When washing dusty clothing, bedding, curtains, etc, add 3-5 drops of neem essential oil to the rinse cycle.

Get Your FREE Bonuses Today!

FREE Fertility Advice from 'The Bringer of Babies'

Leading natural fertility specialist, Gabriela Rosa (aka The Bringer of Babies) has a gift for you. As a thank you for purchasing this book get your FREE "Natural Fertility Booster" subscription and discover...

- Easy ways to comprehensively boost your fertility and conceive naturally, even for women over 40;
- Natural methods to dramatically increase your chances of creating a baby through assisted reproductive technologies such as IUI, IVF, GIFT or ICSI;
- Simple strategies to help you take home a healthier baby;
- How to prevent miscarriages.

You will also receive the FREE audio CD "11 Proven Steps To Create The Pregnancy You Desire And Take Home The Healthy Baby of Your Dreams" a total value of $397!

Claim your bonuses at

www.NaturalFertilityBoost.com

Be quick, this is a limited offer.

(Your free subscription code is: PYF)

FLEAS

About the Flea

Fleas are seldom a problem in homes without pets. The main way fleas enter the home is by being carried in by the family pet, such a dog or cat. Once inside, fleas can become dislodged from the host animal by sudden movements. They may either jump back on the pet or bury themselves in floor cracks or carpets. Once fleas have established themselves in carpets or pet bedding, disturbing an infested area is likely to cause them to jump up and bite your foot or ankle. Cat and dog fleas are very common the world over, if you discover fleas in your home, this is very likely the kind you have. While they normally infest dogs and cats, they feed on almost all warm-blooded animals.

Fleas are wingless insects with very dark reddish-brown, almost black, flat bodies which are compressed from side to side to allow them to move easily between hairs. Their mouth parts are designed for piercing and sucking, as the adult flea feeds on blood. They have short, stout antennae, which are set in indents just behind the eyes.

A flea's life cycle is a complete metamorphosis and it can go from egg to adult in the space of two weeks. The adult female flea lays tiny whitish eggs in the fur of the infested animal. The eggs are very loosely attached to the fur and readily fall to the floor when the animal lies down, or moves. Under favourable conditions, the eggs hatch in about two to four days into long slender white worm-like larva. These larvae feed on bits of dust, food crumbs, and flea faeces and will crawl into cracks and crevices to find accumulations of dust. The larva will shed its skin about two times before spinning a silk-like cocoon. Inside the cocoon the pupa is formed and under ideal conditions the flea remains in the pupa state for about four days before hatching out as an adult. If there are no hosts nearby, a flea cocoon can lay dormant up for up to a year, or until activity is sensed. This is why a house which has been left vacant for a long period of time may suddenly have an active flea population.

Fleas are possibly the best jumpers in nature, reportedly jumping over 200 times their own height. They can also carry dangerous and sometimes fatal diseases. The common flea has been known to transmit tapeworm and is one of the main transmitters of bubonic plague. This is why it is of the utmost importance that flea infestations be addressed immediately, so they do not threat to the health and wellbeing of you and your family.

A flea bite is uncomfortable, and in most cases causes swelling and intense itching. However, the discomfort is usually of short duration. If you are experiencing an adverse reaction to a flea bite, you should consult a health professional immediately.

Controlling Fleas

Pets bring fleas in to the home, but unless you have only discovered a few fleas on your pet after a walk, the source of the problem is usually a flea infestation in the garden. For a comprehensive flea control program, you will need to treat the garden area, inside the home, and your pets. If one of these areas is neglected it is likely the flea infestation will continue. If you eliminate all the fleas inside the home, the next time your pet goes outside, he will bring more fleas in, or if you get all the fleas off your pet, the next time he lies in his bed, fleas will jump on again.

If your home has been infested it may take several weeks before the infestation is eliminated. It takes persistence to completely stop the life cycle of the flea. If any eggs are left to hatch, the cycle can begin all over again.

Note: Factors such as humidity and the extent of the infestation will cause treatment lengths to vary. A small infestation may be eradicated after one treatment while a larger one may take several treatments over a few months.

Tips and Hints for Flea Control

This is very important—don't close off a room with a flea infestation! If you have a pet let him wander about as normal, in fact, normal household activity is extremely important. Shutting off a room, or decreasing activity will only delay the control program, as the fleas won't hatch out again until the room is reopened. Instead, the flea cocoons will lay dormant until activity is sensed and the infestation will reoccur.

The more activity the better, frequent vacuuming of carpets will agitate the flea pupae, making sure all of them hatch.

THREE STEP FLEA CONTROL PROGRAM

Step 1: Pet Treatment

Your pet is an important member of the family and should never have to suffer from painful, irritating flea bites. In the pet section, later in this book, there are recipes for natural shampoos and pest repellents. As another option, there have been many lines of natural pet products, such as herbal shampoos, supplements, conditioners, and flea collars, introduced in recent years. For a list of companies that focus on a naturally healthy pet, please see the resources guide at the end of this book.

Flea Comb

Begin by thoroughly combing your pet using a flea comb. If you can, place your pet in a bath with water that covers him up to the neck. Fleas can drown in water, though it usually takes a few minutes, so to avoid drowning, they will migrate to the highest safe area, which will be your pet's head. You can then comb the fleas from your pet's muzzle and head. Don't forget to check ears as fleas may try and hide there as well. Collect fleas and place them in a plastic jar or sealable plastic bag. Then freeze the bag or jar before disposing of the fleas.

Drain the bath water. Wash your pet with a flea preventive shampoo, such as one from the pet section of this book. Dry your pet thoroughly using a hair dryer and comb again with a flea comb to make sure you have collected all the fleas.

Hint: Dip the flea comb in alcohol between uses.

Mist your pet's bed with a flea mist, such as Pennyroyal Flea Mist (see flea control recipes). Wash the bed in the hottest water allowed by the manufacturer. Follow by misting the floor under your pet's bed, and any areas frequented by your pet, such as around pet food dishes etc., with a flea control mist.

Before your pet goes outside, spray him with a natural flea repellent (see pet pest control recipes).

Once you have removed all the fleas from your pet, it is important for on-going control, to keep your pet free of fleas. Do this by bathing your pet frequently, about once a week, and combing for fleas every day, especially if you are experiencing flea activity. It is also a good idea to check your pet occasionally for fleas, even if you haven't noticed any, to allow for speedy intervention.

A great natural flea repellent for pets is brewers' yeast–tablets, usually available at health food stores. Feed pets one tablet for every 4.5 kilos (10 pounds) they weigh, for example, a 4.5 kilo (10 pound) dog would get one tablet while a 16 kilo (35 pound) dog would get 3½ tablets. Brewers' yeast causes pets to give off an odour that is unpleasant to fleas.

Step 2: Indoor treatment

If you have managed to catch any fleas before they become an indoor infestation, skip this step; however thoroughly check for fleas before moving on.

Firstly, find the infested areas of your home, which can be any room, especially those with carpets. A hands on, or perhaps I should say feet on, method, is to walk all over your carpet, especially low traffic areas, wearing high white socks. Fleas will jump up when disturbed and you should be able to see them against the white background of the socks, which will also help protect your ankles from flea bites.

Walking across the room like this should bring fleas that have been hiding deep in the fibres to the surface and can cause those in cocoons to hatch, so this is an optimum time to vacuum and catch those active fleas. Vacuuming is an effective way of controlling fleas and should be done daily during infestations.

Once you have vacuumed your carpet apply a carpet treatment powder such as Flea Control or Flea Sneeze Powder (see flea control recipes) over your entire carpet, allowing the power to sit for at least three hours (24 hours would be optimum), then vacuum again and reapply the power. Keep vacuuming daily and applying treatment powders until you no longer see fleas when you walk across the carpet.

Rugs can be rolled up, taken outdoors, and hung in the sun for a few days. They can be treated with carpet powders outside. While the rugs are up, wash the floors thoroughly with soap and water, and dry well.

Upholstered furniture can be treated in the same way as carpets.

Washing pet bedding frequently in hot water will help reduce the chances of infestation or re-infestation, be sure to wash the floor under the bedding with soap and water.

During a severe indoor flea infestation, having your carpets steam cleaned can be a worthwhile investment. Ideally, it should be done about once week for one month and be sure to vacuum thoroughly after each steam cleaning.

Step 3: Garden Treatment

One of the best ways to prevent fleas in your garden and consequently in your home, is by applying an outdoor treatment before flea season begins.

To treat the outside of your home, mix up a natural soap and essential oil solution, one with rose oil or pennyroyal oil works especially well against fleas (see flea control recipes). Spray the mixture around the outside of your home and in areas where you suspect flea activity, also spray the mixture on the ground under plants, however be careful not to spray the actual plants. Another option is a natural citrus-based cleaner. Mix approximately 30ml of citrus cleaner per four litres of water.

Beneficial nematodes may also be applied to the soil to help control flea larvae as well as grubs and other lawn pests. And you may need to try several recipes before you find the one that best suits your needs.

FLEA CONTROL RECIPES

MAKE THE 'FLEA'S SNEEZE'

You can depend on this formula.

Ingredients:

- **3 cups bicarbonate of soda**
- **10 drops of sweet orange essential oil**
- **10 drops of citronella essential oil**
- **8 drops of either peppermint or spearmint essential oil**
- **6 drops lemon balm essential oil**

Yields: 3 cups

Time to make: About 5 minutes
Shelf life: Indefinite
Storage: Glass jar

Method:

Combine bicarbonate of soda, sweet orange, citronella, the mint, and lemon balm essential oils in a large bowl and whisk together.

How to use:

Spread over all your carpets and let stand for a minimum of an hour, preferably overnight when possible. Then vacuum thoroughly.

Note:

Keep separate bowls, and mixing tools for your herbal solutions. Do not use the same tools for cooking. I keep my herbal bowls, mixing tools and the like with my other gardening tools, so as to never confuse them. You may also choose a special colour for your herbal tools, but still do not store them with your regular kitchen tools.

You should dispose of the vacuum cleaner bag once you are finished vacuuming, as this will help prevent future infestations. If you have a 'bag-less' vacuum cleaner, (a better choice for the environment, by the way) take the filter and collection cups apart and wash in hot water, placing a little of this mixture straight into the cup prior to re-assembling.

FLEA CONTROL POWDER FOR CARPET AND UPHOLSTERED FURNITURE

You can sprinkle this powder on carpets or furniture to kill fleas.

Ingredients:

- **1 cup bicarbonate of soda**
- **1 cup cornstarch**
- **2 cups diatomaceous earth (natural, not pool grade)**

Yields: 4 cups

Time to make: About 5 minutes
Shelf life: Indefinite
Storage: Screw top glass jar

Method:

In a glass jar combine bicarbonate of soda, cornstarch, and natural diatomaceous earth. Secure lid tightly and shake to mix.

How to use:

Sprinkle liberally onto carpets or furniture, and allow sitting for about three hours, vacuum thoroughly.

CITRUS FLEA MIST

Citrus is one of the best natural flea repellents as it can kill eggs, larvae, pupae, and adult fleas. This is a very effective mist for controlling fleas in carpets, pet bedding, etc.

Ingredients:

- **4 cups water**
- **4 teaspoons orange essential oil or citrus peel extract**

Yields: 4 cups

Time to make: About 5 minutes
Shelf life: Indefinite
Storage: Glass jar, spray bottle

Method:

Combine water and orange essential oil in a glass jar, secure lid, and shake vigorously to blend. Pour half the mixture into a spray bottle and store the other half in the glass jar.

How to use:

Mist onto infested pet bedding, carpet, and other infested areas.

MINTY FLEA MIST

This mist contains peppermint castile soap and is very effective against fleas.

Ingredients:

4 cups water
4 teaspoons peppermint castile soap

Yields: 4 cups

Time to make: About 5 minutes
Shelf life: Indefinite
Storage: Glass jar, spray bottle

Method:

Combine water and peppermint soap in a glass jar, secure lid, and shake well to blend. Allow the suds to subside, then pour half the mixture into a spray bottle and store the other half in the glass jar.

How to use:

Mist onto infested pet bedding, carpet, other infested areas, etc.

PENNYROYAL FLEA MIST

Pennyroyal has been known since antiquity as an insecticide. The original French name meant “a flea”.

Ingredients:

- **4 cups water**
- **2 tsp. Pennyroyal essential oil**
- **2 tsp. Eucalyptus essential oil**

Yields: 4 cups

Time to make: About 10 minutes
Shelf life: Indefinite
Storage: Glass jar, spray bottle

Method:

Combine water, pennyroyal and eucalyptus essential oils, in a glass jar. Secure lid, and shake well to blend. Pour half the mixture into a spray bottle and store the other half in the glass jar.

How to use:

Mist onto infested pet bedding, carpet, other infested areas, etc.

Note: Pennyroyal is contraindicated in pregnancy. Consult your health care professional or herbalist before using. Keep out of the reach of children and pets to avoid accidental ingestion.

ONION FLEA MIST

This recipe needs a little extra time and cooking to make, but your effort will be rewarded with an excellent flea repellent.

Ingredients:

- **4 cups water**
- **2 large onions or 4 small onions**

Yields: 4 cups

Time to make: About 1 hour
Shelf life: Indefinite
Storage: Glass jar, spray bottle

Method:

Peel and slice the onions and place them in a large pot. Add four cups of water, or enough to completely cover the onions. Bring to a boil, reduce heat and simmer.

Check frequently for evaporation and add water if needed to keep the onions covered. After 30 minutes remove from heat and cool. Remove the onions and strain mixture. Pour half the mixture into a spray bottle and store the other half in a glass jar with a lid.

How to use:

Mist onto infested pet bedding, carpet, other infested areas, etc.

Tips and Hints for Flea Control

- Fleas despise any strong breakfast tea. Tear open a few bags, scatter the loose tea about on your carpet and vacuum up in a few days.
- At night, place a dish of soapy water near areas with flea activity and hang a light bulb over the dish. The warmth of the light bulb will attract the fleas causing them to jump into the dish of soapy water.
- Make sure all you windows have screens and doors and windows have snug seals.
- Rodents have long been a contributing factor in flea infestations. Eliminating unwanted rodent populations will reduce fleas and also reduce the chance of contracting infectious diseases. Rodent fleas are one of the main carriers of bubonic plague and while cases are rare, they do still occur.
- You may want to consider a cat as a pet, as their natural hunting instincts make them the perfect natural rodent control, which in turn will reduce flea infestations.
- Diatomaceous earth works well against fleas. Lightly sprinkle this around the inside or outside of your home at the beginning of flea season.

Gabriela Rosa

FLIES

About the Common House Fly

The common house fly (Musca domestica) is a pest common to all areas of the globe. It belongs to the insect order Diptera, meaning two-winged. Flies prefer to lay their eggs in horse manure but will also breed in other types of excrement or rotting vegetables. Female flies lay clusters of about 120 to 150 white, elongated eggs, which look like grains of wheat. Depending on environmental conditions, these eggs hatch about six to eight hours after being laid. During the summer months, when house flies are most active, the complete life cycle of a fly, from egg to adult can occur in 10 to 14 days.

Active and agile, fly larvae are called maggots and are whitish in colour. Maggots moult three times before they change into pupae. The last skins of the maggots remain on the pupae so they are encased within hard dark brown protective shells. After their last moult, pupae rest for about five days, or for as long as an entire winter, until the adult flies emerge from the protective cases.

The adult house fly is about 5 mm in length and has thin veined membranous wings. The body is greyish in colour with four darker grey strips running around the thorax. The adult fly's body and legs are covered with tiny hairs and bristles that collect large amounts of germs and bacteria as the fly goes from place to place. Each foot ends with two sticky, hairy pads that help the fly climb up walls and defy gravity as it walks across the ceiling. These pads excrete minute drops of fluid, which literally make the fly stick to the wall, and they also collect all kinds of bacteria from the filthy things the fly walks on. Wherever a fly lands it spreads millions of bacteria and germs it has collected from unsanitary surfaces. This is why house-flies are such a health concern.

The housefly does not bite. Its mouth parts are not suited for piercing, but for sucking and lapping up liquids. Flies eat solid substances, such as sugar, by first moistening and dissolving them with saliva before drinking them up.

Controlling Flies

With our advanced waste management systems, we do not suffer from the enormous fly populations of the past. However, flies are still a concern, especially during the hot summer months when they are especially active, because they carry and transfer dangerous diseases such as typhoid, anthrax, cholera, tuberculosis, dysentery and many others.

One of the most effective ways to control flies is to remove any fly breeding grounds, such as compost heaps or uncovered garbage, from near your house. Studies have shown that while flies do not normally travel far from their breeding places, their flights are very susceptible to wind currents and these can help them travel a kilometre and a half (a mile) or further from their original breeding place.

Natural Enemies of the House Fly

In nature, house-flies have many enemies. Some are plants, such as fungi; however, their main enemies come from the animal kingdom. Spiders are a natural predator and will help reduce fly numbers if allowed to build their webs in your garden. Hornets are not as effective as spiders at catching and killing flies, however they are natural predators and do help a little. There are also some kinds of mites that have been observed attaching themselves to flies. However, they seem to do this as a mode of transport and when the flies land near a source of food, the mites detach themselves and fall off. Yet another excellent reason for fly control!

Flies love filth, one of the best ways to prevent flies and maggots in your home is to keep it perfectly clean. Make sure all kitchen scraps are thrown away and counters are wiped clean. It is also a good idea to empty rubbish bins at the end of each day and remove the garbage to an outside location.

Flies are drawn to compost bins, so make sure any compost containers are tightly sealed. Never put meat scraps in the compost bin, as flies are particularly attracted to rotting meat.

One of the best ways to prevent flies from entering your home is to make sure all windows and doors are screened or closed. Place fly deterrents such as mint, basil, or cotton balls treated with pine oil near windows. You can also wipe windows and door frames with a clean cotton towel moistened with water and a few drops of pine essential oil.

Hang a few fly paper traps around your home to capture indoor intruders. See the fly control recipe section for some natural homemade fly papers.

If you keep your home clean and your outside garbage in good order you should have little trouble with flies or maggots.

Don't be discouraged if one repellent doesn't work. Each insect reacts differently to the natural repellents. You may need to try several fly repellents before finding the one which works best.

HOUSE FLY CONTROL RECIPES

FLY NO MORE PAPER

You can make your own flypaper with this simple recipe.

Ingredients:

- **¼ cup maple syrup**
- **1 tbsp. granulated sugar**
- **1 tbsp. brown sugar**
- **Brown craft paper such as a brown paper bag**

Yields: Several fly paper traps

Time to make: About 20 minutes
Shelf life: Use immediately
Storage: N/A

Method:

In a small bowl combine syrup, granulated sugar, and brown sugar and mix. Use scissors to cut strips of the brown craft paper and soak them in the mixture. Lay the strips of paper out flat and allow them to dry overnight.

How to use:

Make a hole on one end of a strip. Choose a location where the sticky strip can hang freely and undisturbed, but in an area frequented by flies. Cut a piece of thread or twine to the length necessary to hang the strip, loop it through the hole and tie it. Position the strip and tie or tape the string where you want the strip to hang. The flies are attracted by the sweetness, and become stuck to the sticky paper. Replace strip as needed.

FLY AWAY HERBAL SACHET

These not only repel flies but also smell great!

Ingredients:

- **2 Parts Clover blossoms**
- **2 Parts Eucalyptus leaves**
- **1 Part Cloves**

Yields: Several sachets

Time to make: A few hours
Shelf life: Until the fragrance disappears
Storage: Sachets

Method:

In a glass bowl, combine clover blossoms, eucalyptus leaves and lightly crushed cloves. Gently toss to mix.

How to use:

Spoon into a small bag (a wedding favour bag works well), or make your own by sewing a small square from cotton cloth. Sew the edge closed or pull closed the draw string and hang the finished sachet by an open door or window.

These are especially nice in the spring when doors and windows are frequently left open, allowing you to enjoy the fresh new scents of the season.

BASIL FLY MIST

Basil is especially effective against fruit flies as you'll discover when you use this great mist.

Ingredients:

4 cups water
4 teaspoons of basil essential oil

Yields: 4 cups

Time to make: About 5 minutes
Shelf life: Indefinite
Storage: Spray bottle or glass jar

Method:

Combine water and basil essential oil in a glass jar, secure lid, and shake vigorously to blend. Pour half the mixture into a spray bottle and store the other half in the glass jar.

How to use:

Mist around window and door frames to repel flies, or spray on active fruit flies.

Tips and Hints for House Fly Control

- A few drops of eucalyptus oil on a piece of absorbent cloth will deter flies. Leave in areas where flies are a problem.
- Small sachets of crushed mint can be placed around the home to discourage flies.
- Safe, non-toxic, pheromone-based fly traps are available for both indoor and outdoor use.

- Pour boiling water with 20 drops of citrus essential oil down drains to eliminate nuisance gnats and flies that may be attracted to food odours.
- Place a good handful of dried cloves in a bowl and coat with four to six drops of clove, lavender, and citronella or peppermint essential oil. Freshen with additional oil as needed.
- Herbs which work well against flies are basil, cloves, and pine essential oil.
- Remove any pet faeces from you garden. If you have a pet, you should also have a special shovel and tightly sealing, lined rubbish bin for the easy disposal of pet faeces.
- Rubbish bins with tightly fitting lids will reduce fly breeding grounds. Keep the bins tightly closed.
- Consider planting basil by your kitchen door or window, or plant some in a window box or indoor herb garden.
- Clean garbage bins every few weeks with boiling water and peppermint castile soap.
- A wine bottle with a few fruit peels in the bottom will lure flies in, however the flies can't fly out of the narrow bottle neck.
- Planting marigolds around your yard works as a natural bug repellent because the flowers give off a fragrance flies do not like.

MOSQUITOES

About the Mosquito

Mosquitoes belong in the fly family and only differ from the common house fly in size and ability to bite. The hum of this insect has a very ominous meaning, as some mosquitoes are the carriers of dangerous diseases. One species carries yellow fever, and other two carry malaria. One common genus, Culex, carries several types of encephalitis, and West Nile virus. In Australia, important human diseases transmitted by mosquitoes include Dengue fever, Australian encephalitis, Ross River virus disease and Barmah Forest virus disease. While dangerous species of mosquito do exist in large numbers, the most common mosquitoes are not carriers of these diseases.

Mosquitoes breed in both fresh and salt water. The largest infestations seem to be in coastal regions, which would indicate that salt water is the best breeding ground. However, common mosquitoes breed anywhere fresh water is present; this includes still pools, ditches, garbage cans, old tyres, ponds, etc.

In autumn, adult mosquitoes hideaway in basements or other places suitable for their dormant condition in winter. Most of the hibernating mosquitoes die before spring. However, the female mosquitoes, which do live through the winter, become active in spring and immediately search out a place to lay their eggs.

Mosquito eggs are deposited on the surface of the water in a shape resembling a boat. Each of these egg masses contains between 75 and 200 eggs. The eggs are easily visible to the eye and are from 3mm to 6mm in length. Depending on the water temperature, the eggs will float on the water's surface for 24 to 76 hours before hatching.

You have probably seen little 'wrigglers' in stale or stagnant water, these are mosquito larvae. They hatch from the lower portion of the egg into the water. The larva has a repertory tube at the tip of its abdomen, which it rests at the surface of the water to draw in air. On either side of its mouth there are dark hairs, which vibrate constantly, creating small currents that draw food near the mouth. In about five to ten days the larva changes into a pupa.

The pupal stage is often confused with the larval stage because the pupae are such active wigglers. The pupa has a slender abdomen, with a large head and thorax. It has two repertory tubes in its thorax. In about five or six days the pupa transforms into an adult. The adult mosquito emerges through a slit along the top of the thorax and rests for a few minutes on its shed skin while it dries its wings before flying off.

The adult female will live two to three weeks while she searches for a place to deposit her eggs. During this interval the female mosquito may bite several times. The male mosquito however is usually not annoying and does not bite. It also never takes blood meals, only females do that, and only at a certain point in their life cycle when they need iron for the eggs they are about to lay. Both males and females mainly eat plant materials such as nectar. Male and female mosquitoes are distinguished by their antennae. Females have a few short hairs on their antennae, while male antennae are covered with many long bristles making them look slightly like feathers.

The Mosquito Bite

Mosquitoes are equipped with a long beak made up of six bristle like organs enclosed within a sheath. The long slender beak is not the organ which punctures the skin. The sheath bends to reveal the bristles that extend past the end of the beak and bore into the flesh. While the mosquito bites it injects poisonous saliva into the wound, this is what causes the itching and swelling from mosquito bites.

Controlling Mosquitoes

Adult mosquitoes are very difficult to destroy so it is best to focus on the egg, larva, and pupa stages. One the best ways to eliminate the immature stages of the mosquito is by draining any stagnant bodies of water that could harbour them. Empty any tanks, tyres, buckets, or cans of standing water and consider installing drainage in any area that has a tendency to fill with water.

Often there are some bodies of water or pools that can't be drained for one reason or another. Ponds that have been installed for enjoyment greatly add to the beauty of the landscape, however they do provide an ideal breeding environment for mosquitoes. Fish are natural predators of mosquito larva, so consider adding a few fish to your pond.

Adding oil to stagnate water, such as in a rain barrel, is another way of preventing mosquitoes breeding in these locations. One tablespoon of olive oil on the water's surface should be enough to prevent wrigglers.

Plants for Your Home and Patio that Prevent Mosquitoes

- Basil planted close to your home will repel mosquitoes.
- Lavender planted all around your home will repel flies and mosquitoes.
- The castor bean plant is terrific as a house plant. Keep in pots indoors to repel mosquitoes, and as they outgrow the pots, replant them around your patio to repel mosquitoes, and begin new plants indoors.
- Scented geraniums can be planted in pots indoors and out to repel mosquitoes.
- Citrosa plants are useful indoors in pots and out in these are also great for this purpose.
- Lemon thyme can be planted in pots on your patio.
- Citronella grass can be nurtured anywhere you wish to repel mosquitoes.
- You can also use common marigold in the house and outdoors as a repellent.
- Thai lemon grass is an effective mosquito repellent, and well worth cultivating also for the beautiful flavour it can add to your food.

Tips and Hints for Mosquito Control

- Prevent annoying mosquitoes at a summer barbeque by placing a few sprigs of rosemary or sage on the coals.
- Burn lemongrass and citronella essential oils indoors and outside place a bowl of water with several drops of the same essential oils to help keep mosquitoes away.
- Fill a 500ml spray bottle with water, add about 10 drops (in total) your favourite mosquito repelling essential oils and spray indoors and out.
- Create a new design element and mosquito repellent for your doors and windows by hanging a ribbon treated with 6-10 drops of either peppermint, lavender, citronella, or lemongrass essential oil.
- An open bottle of pennyroyal essential oil can help keep mosquitoes away.
- Chamomile and citriodora are also great essential oils for repelling mosquitoes.
- Mosquitoes can also be repelled with neem essential oil, which can be put inside a spray bottle with water and sprayed around the house.
- Fine netting or screens will help keep mosquitoes out.

MICE

About Mice

Mice, along with their cousins the black and brown rat, which are now so abundant all over the world, originated in Asia. The early explorers introduced these rodents into countries such as Europe and America.

Adult mice are typically very light, small creatures with long tails. They have very strong hind legs that enable them to stand erect and even walk a few steps in an upright posture. They are natural swimmers and have a very well developed sense of hearing and taste. Their diet is varied but they thrive on vegetables, fruits, nuts, grains, and pet food.

During a year, a female mouse will produce between four and six broods of four to eight young, making it a very prolific breeder. The mother mouse builds a nest from bits of straw, shredded paper, feathers, wool and other soft materials. When they are born, baby mice are tiny, blind, hairless, nearly transparent helpless creatures. They grow rapidly and are completely covered with fur within one week but don't open their eyes until they are about two weeks old. A couple of days after they have opened their eyes they will leave their nests and begin foraging for food on their own.

Mice droppings are sometimes confused with cockroach droppings, as they look similar, however they are slightly larger, being about 3mm long, dark coloured and capsule shaped. The average life expectancy for a mouse is about one year.

As the seasons change from autumn to winter, mice are especially aggressive about entering homes in search of new food supplies and warm places to spend the cold months. Mice and other rodents will chew through storage containers and contaminate human food. It is a good idea to protect your food supplies including crackers, bread, any grains, nuts, fruits, vegetables, pet food, sugar, flour, etc in tightly sealing canisters such as glass canning jars and heavy plastic containers, or keep them in the refrigerator.

Mice, like other fur covered animals, shed their fur and can contaminate food with hairs, as well as droppings and urine. Mice and rodent droppings however are of paramount concern because they transmit diseases such as leptospirosis, murine typhus, rat bite fever, salmonellosis, Seoul virus, and trichinosis.

Birds of a Feather...

Cockroaches and mice often coexist happily in the same habitats. They are both nocturnal and live in dark, hidden locations. The cockroaches eat the mouse faeces and the mice, in turn, may eat the cockroaches. Cockroaches are able to eat commercial mouse baits such as pellets and blocks with no adverse results. If you are using commercially produced bait to control mice, you may be inadvertently feeding the cockroaches. Which will, in turn, be food for the mice...

Mice will also aid other pantry pests in their search for food, since they sometimes collect seeds or dry pet food and hoard it in the walls, or under cupboards or dishwashers where the source of infestations can be nearly impossible to find.

Controlling Mice

Mice love to make their homes in warm, dark areas, such as walls with hot water pipes, under kitchen counters, in closets, basements, crawlspaces and other warm, dry quiet places. Because mice are so resilient, they can easily become a major nuisance. Once they become established, you will need a proactive trap and bait program and perfect cleanliness to eliminate the infestation.

Tips for Mice Prevention

- Make sure all kitchen scraps are thrown away and kitchen benches are wiped clean. It is also a good idea to empty rubbish bins at the end of the each day and remove the garbage to an outside location.
- Mice are drawn to compost, so make sure any compost containers are tightly sealed.
- One of the best ways to prevent mice from entering your home is to eliminate or plug any cracks, holes or entrances leading in from the outside. Look around your home for any places mice are likely to enter. Good places to check are around the foundations, electrical boxes, and door or window frames. Mice, especially young ones, can fit through very small holes, so caulking any cracks and ensuring door thresholds are snug will reduce the chances of them entering your home. You can also add a few drops of peppermint, eucalyptus, or spearmint essential oil to a cotton wool ball and place anywhere you suspect rodents may be invading your home.
- Place mice deterrents such as mint, eucalyptus, or cotton balls treated with peppermint, lavender, or spearmint essential oil behind kitchen appliances, behind furniture, in drawers, in shelves, in closets, and in the pantry.

Don't be discouraged if one repellent doesn't work. Each mouse reacts differently to natural barriers. You may need to try several mice deterrents before finding the one that works best.

TWO STEP RODENT CONTROL PROGRAM

This comprehensive program will help eliminate mice and other rodents both indoors and out, while reducing the chance of re-infestation.

Step 1: Inside Treatment

Setting baited traps inside your home works especially well because mice and other rodents are foragers and are constantly searching for food. Sometimes the only option is to use a traditional spring mouse trap although they can be messy and are not always humane. If you can get hold of them, humane mouse traps that catch the mouse in a box for subsequent release away from the house may suit you better. However, bear in mind that the mouse's chance of survival may be slim when out of its territory and up against cats, birds and other predators. See resources section for more information about live mouse traps.

A few good baits for mice include, cheese, bacon, bread and butter, small nuts, oatmeal, sunflower, seeds and other grains. A mixture of peanut butter and oatmeal also works well. Poisoned baits for rodents are not recommended because it makes them sick, and easy for family pets to catch. The pets are then poisoned by eating the poisoned rodent.

Set as many baited traps around your home as needed. Good locations include behind furniture, behind kitchen appliances, under sinks, and anywhere you have noticed rodent activity. You may need to experiment with a few different baits at first, to find the one which best suits your mice.

Check your traps daily and reset and re-bait them if necessary. If a trap is receiving a lot of traffic, consider placing another trap nearby to catch even more mice or rodents in the area. It is a good idea to keep a few baited traps around your home (away from pets and children), even if you are not experiencing any rodent problems. These traps only need to be checked every week to make sure they haven't been tripped.

If you haven't seen any mice but suspect you might have them, inspect your home for signs of rodents. Look for droppings, gnawed food packages, and oily marks along walls where the greasy fur of the rodent has been rubbing. The most likely areas of mice activity are the kitchen, pantry, and other areas where food is stored.

If you find a rodent nest, do not disturb it. Disturbing it will only cause the mouse or rodent to move to another location in your house, instead, set a trap near the nest. You will probably have to set it several times to capture all the occupants of the nest.

Note: Baiting works best when it is the mouse's only food source. Make sure all garbage is removed, counters and floors are free of crumbs, and all food packages are sealed tightly.

Step 2: Outdoor Treatment

Often mice enter your home from breeding places such as unmaintained garbage areas, or nearby woods or fields. Creating an outdoor barrier may stop mice before they enter your home. Some of the best outdoor barriers are herbs. Planting basil, mints, and lavender around the perimeter of your home will repel mice and rodents.

Check the outside your home for any small cracks or places, such as around electrical boxes or garage doors, where mice might gain entrance. Make sure there aren't any holes in easily forgotten places such as roof spaces, attics or basements. Caulk or plug any cracks or holes you find.

The most important element in rodent control is cleanliness and proper storage of garbage. Make sure all garbage bins have tightly fitting lids and the area around them is clean. Eliminate potential mouse breeding grounds by keeping the garden tidy and getting rid of any scrap heaps. Wood piles should be as far away from the house as possible.

MICE CONTROL RECIPE

MICE MIST

Peppermint and spearmint essential oils both have a scent that pleases humans but that rodents detest. Drive the pests away with this refreshing mist...

Ingredients:

4 cups water
1 tsp of Lavender essential oil
1 tsp of Peppermint essential oil
1 tsp of Spearmint essential oil
1 tsp of Eucalyptus essential oil

Yields: 4 cups

Time to make: About 5 minutes
Shelf life: Indefinite
Storage: Spray bottle and glass jar with tightly fitting lid

Method:

Combine water and your essential oil in a glass jar, secure lid, and shake vigorously to blend. Pour half of mixture into a spray bottle and store the other half in the glass jar.

How to use:

Mist around window and door frames to repel mice. Spray along walls, or in active areas of known mice activity. Re-spray daily or as often as needed.

Note: Be sure to test in an inconspicuous spot before misting on walls or flooring.

Tips and Hints for Mice and Rodent Control

- Placing two or three drops of essential oil on to cotton balls and locating them in problem areas is a great method. Add more oil for a stronger scent and refresh as the oils lose their potency. They will need to be replaced every few days.
- A combination of cotton balls and sprays works well. Drop the cotton balls behind stoves and other hard to reach places, and use Mice Mist freely elsewhere.
- Citrus is also offensive to rodents, so polishing your wood cabinets with a nice orange based oil will also annoy them.
- Finally, soft drinks don't only destroy our health, they actually kill rodents. Mice are attracted to soft drinks because of the sugar, but because they cannot burp, they die soon after drinking it. If using it for this purpose be sure to replace the soft drink daily as it loses carbonation within several hours.

CLOTHES MOTHS

About the Clothes Moth

Clothes moths have been with us for centuries, they were familiar pests to the ancient Romans, who called them Tinea, prompting modern scientist to name this family of moths Tineidae. It's not the moth, but the larvae of the clothes moth that damages clothing. The larvae live on animal products such as wool, leather, other dead insect bodies, feathers, hair, fur, etc. The have followed man throughout history, feeding on primitive furs right through to fine wool and silk.

The typical clothes moth is small and slender and can be quite beautiful. They are about 12mm wide when the wings are fully extended. The forewings are a luminescent brownish-yellow colour with dark spots on each wing. Clothes moths normally appear in the spring months and you will probably see them flitting about your home until the summer months. It is a good idea to remove these moths when you find them to prevent them from laying eggs that turn into the larvae that are the real source of destruction. Female clothes moths lay their tiny eggs in the folds of clothes or other articles, where the larvae hatch out and immediately start feeding.

Controlling Clothes Moths

Clothes moths love the darkness of wardrobes, drawers, and storage containers, so one of the best ways to control infestations is to remove all the affected items from your wardrobe and let them air outside in a hot sunny location for about two to three days. Clothes moths are very susceptible to heat so the hotter the day the better.

If clothes can't be placed outdoors, vigorously shake them over a rubbish bag then closely examine the article and remove any larva which didn't fall off.

To repel clothes moths, place a cotton ball which has been treated with cedar wood essential oil in wardrobes, drawers, and storages bins. Before storing clothes for extended periods, examine them for any moths, larvae, or eggs. One accidentally trapped female moth can ruin any attempts to protect your clothes.

CLOTHES MOTH CONTROL RECIPES

MOTH AWAY SACHETS

This is a wonderful alternative to mothballs. The main ingredient in mothballs is either naphthalene or paradichlorobenzene, both of which are dangerous chemicals that cause symptoms such as nausea, headache, swollen eyes, irritation to eyes, nose and throat, and lungs, as well as depression.

It is disturbing that the labels say to avoid inhalation of vapours! This is not so possible since they are placed in the wardrobe and the smell permeates everything inside including the room. Natural alternatives are the best remedy.

Ingredients:

- **4 tbsp dried lavender**
- **2 tbsp dried orange peels**
- **2 tbsp cedar chips**
- **1 clean white cotton sock**

Yields: ½ cup

Time to make: About 5 minutes
Shelf life: Indefinite
Storage: Air tight container

Method:

Put the dried lavender, cedar chips, and dried orange peel in the sock

How to use:

Tie with a string and hang in a wardrobe or place in a drawer or chest. It keeps the bugs at bay and your clothes will smell lovely.

Variation:

Instead of lavender, cedar, and orange, combine 1 ounce each dried ginseng and thyme and 2 ounces each dried peppermint and rosemary.

And You Can Create Your Own!

There are some great moth repelling herbs which you can mix, and combine into personalised sachets. Because they hang in your wardrobe, your clothes will absorb some of the fragrance you have created. Some great moth repelling herbs you can use:

- Cedar
- Cloves
- Lavender
- Mints
- Hyssop
- Tansy
- Thyme
- Rose
- Rosemary

VINTAGE MOTH REPELLENT

Natural camphor oil is a strong volatile oil that has been well known for centuries for its moth repelling qualities.

Ingredients:

3 cups water
1½ tsp camphor oil (be sure to purchase natural camphor oil only. Do not use synthetic versions)

Yields: 3 cups

Time to make: About 5 minutes
Shelf life: Indefinite
Storage: Spray bottle or glass jar

Method:

Combine water and camphor oil in a glass jar, secure lid, and shake vigorously to blend. Pour half of mixture into a spray bottle and store the other half in the glass jar.

How to use:

Mist into closets, drawers, pantries, storage containers, or anywhere you suspect moths.

Note: Camphor Oil is a very powerful insect repellent and is regulated or banned in some countries. Camphor oil is contraindicated in pregnancy. Consult your health care professional or herbalist before using. Keep out of the reach of children and pets to avoid accidental ingestion.

Tips and Hints for Clothes Moth Control

- Add a few drops of cedar wood essential oil to a cotton ball and place in wardrobes and drawers. An extra drop of spearmint, lavender, citronella, or peppermint essential oil can also be added. Add more oil for a stronger scent and refresh as the oils lose their potency.
- Cedar chips in a cheesecloth square or cedar essential oil in an absorbent cloth will repel moths. The cedar oil must be 'aromatic cedar', also referred to as juniper in some areas.
- Homemade moth-repelling sachets can also be made with lavender, rosemary, vetiver and rose petals.
- Dried lemon peels act as a natural moth deterrent. Simply toss a few into your clothes chest, or tie in cheesecloth and hang in the closet.

MOTHS AND BEETLES FOUND IN FOOD

Preventing pests in your pantry

Troublesome insects are found in grains, flour, cereals, wheat, oats, and rice and also in dried fruit, powdered milk, nuts, popcorn, spices, pet food, macaroni, and crackers, etc. The larvae of moths and beetles and the maggots of some kinds of flies are the main pests found in our pantries. They find their way into our homes with the grains, or cereals we purchase, which, in turn, have been infested at the mills and granaries from which they came.

Controlling Pantry Pests

- Start when you shop. Before you bring home any flour, grains, or cereals, check the packages for any signs of pests. When you get them home, put all pantry items in the freezer for 24 hours to kill any larvae, eggs or insects.
- Keep susceptible foods stored in airtight containers such as glass canning jars.
- To prevent the infestation spreading, foods which have been infested should be thrown away or frozen for one week—this will kill any insects, larvae, or eggs living in the food.
- Wipe pantry or cupboard shelves with a cloth moistened with water and a few drops of peppermint essential oil.
- Crevice and cracks can be hiding places for bugs that are no longer in the food packages. Puff a little diatomaceous earth into any cracks or areas where they could be hiding.
- Vacuuming will also help capture any wandering pantry pests, just be sure to dispose of the vacuum bag when you are finished.

- Even after you have disposed of all infested foods you may still have a few moths or beetles that can hatch out, watch for these and remove them as soon as they appear.
- You can heat infested foods in the oven to 55-60°C for at least an hour to kill all life stages of grain infesting insects.
- Use older packages of susceptible food stuffs before the newer ones. Avoid long storage of cereals, grains, rice, etc.
- Insects that make their homes in our food are not harmful to humans. So, don't worry if you accidentally eat one, or discover a few floating in your milk.

The best way to eradicate these pantry pests is by throwing away the contaminated product and wiping down the shelves and walls of the pantry with a general purpose natural insecticide. To prevent future infestations be sure to implement the pantry pest controls describe previously.

Following are some of the more common pests you will find infesting foods:

Dark Mealworm

These are actually beetle larvae. The dark mealworm (*Tenebrio obscurus* Fabricius) thrives in flour, cake mixes, cereals and bran. The adult beetles are dull black and about 12mm long. As the name suggests the larvae is dark brown.

Yellow Mealworm

The yellow mealworm (*Tenebrio molitor* Linnaeus) is similar to the dark mealworm, but slightly lighter in colour. It is most commonly found in breakfast cereals and macaroni and in feed mills and processing plants.

The adult beetle is a very shiny black, compared to the dark mealworm, which is a dull opaque black. The former deposits its eggs in grains or cereals where they can hatch and begin eating. The eggs are white and can be laid in bunches or singly and are covered in a tacky substance which causes bits of meal to stick to them. The eggs usually hatch into minute white meal-worms in about ten days to two weeks. As soon as they hatch, they begin to eat and gain their glossy yellow appearance. They can remain in the larval stage for as long as 600 days. The pupa stage typically lasts two weeks. In nature, the beetles appear in spring, but in indoor conditions, the beetles can appear at any time. In general, there appears to be one generation per year.

An interesting note about the yellow meal-worm is that while it has been accidentally distributed all over the globe, it was purposely introduced into Chile to help provide food for domestic birds.

Sawtoothed Grain-Beetle

Another bug that infests grains is the sawtoothed grain-beetle (*Oryzaephilus surinamensis),* also known as the grain weevil. This tiny beetle is barely 2.5mm long. Found in all grains, it is one of the most prolific beetles, particularly in warmer regions. The adult is reddish-brown, slender and flat with its most distinctive feature being three saw like ridges on the each side of the thorax (hence its name).

The larva is flattened slightly, and will typically build a light case if it is living in coarsely ground grain such as meal. However, it does not build a case in finely ground material such as wheat. The pupa is whitish in colour. Their life cycle during the summer months is about twenty four days but during colder months, it is considerably longer.

These beetles have a habit of chewing holes through paper bags and often tiny holes in a flour packet or other paper food container is a sure sign of an infestation.

Mediterranean Flour Moth

The Mediterranean flour moth (*Ephestia kuehniella* Zeller) is found in home pantries around the globe. The minute eggs are deposited in flour, pet food, cereals, etc. and as the larvae burrow deep into the products, they easily evade notice as the foodstuffs are transported long distances.

When first laid, the eggs of this moth are nearly white, oval and elongated, however they later become wrinkled and brown. They are minute but visible to the naked eye. The eggs hatch in about five to ten days. When the larva first emerges it is about 1mm long, however it grows quickly and eats a great deal and typically becomes fully grown in approximately 26- 40 days. This process takes longer in the colder months.

The Mediterranean flour moth larva is about 12mm long with a pinkish coloured, cylindrical body and a darker brownish-pink head. The body is lightly covered with long hairs. Once the larva is mature, it spins a silk cocoon, which is usually attached to a surface and often covered with bits of the substance the larva inhabits. It then rests for about ten to fifteen days in the pupa state before it emerges as an adult moth.

The adult moth is about 12-18 mm long when at rest, and 18-25 mm with wings expanded. It has dark blue-grey forewings that have a wavy, "W" shaped design on them. The hind wings are a silvery grey colour. Both wing segments are fringed heavily with long hairs. Like a butterfly, the moth needs to dry its wings before it can fly. The adult female moth can lay as many as 271 eggs in her lifetime. Typically there are four generations of these pests per year, however in warmer climates there may be more.

These moths are partially injurious to food because the larvae crawl thorough the product, such as flour, leaving a trail of spun silk threads wherever they go. If a bag of flour becomes infested with these moths, the entire contents will be filled with their silk webbing, making the product unsalvageable.

Confused Flour Beetle

The confused flour beetle (*Tribolium confusum* Jacquelin du Val) is frequently confused with the rust red flour beetle and the saw-toothed grain beetle. The confused flour beetle is reddish-brown in colour and about 8mm in length, with a flattened oval body. The top side of the body and head are covered in tiny dents.

The larva of this beetle closely resembles that of the yellow meal-worm except it is much smaller, about 6mm length. When the larvae first hatch, they are whitish in colour and gradually turn a hard shiny brown, with lighter coloured bands where the segments of the body join. The larva eats a wide verity of food stuffs including, flour, cereal, cayenne pepper, ginger, peanuts, peas and beans.

It takes about four weeks for the larva to mature and change into the pupa state; it then rests for about six days before emerging as an adult. The entire life cycle of the confused flour beetle takes about thirty-six days.

Indian Meal Moth

The Indian meal moth (*Plodia interpunctella*) is probably the most common insect found in pantries. The larvae are a pale fleshly colour with a sight pink or yellowish tint. The head of the larva is a reddish-brown or yellow. These are often found in corn meal, from which they get the name "Indian meal" moth, however they love to eat any grain, biscuits or dried fruit.

This moth looks like a small Mediterranean meal moth. The upper wings are a white to cream colour at the ends with reddish-brown coloured upper parts with dark grey to black markings and bands. The lower wings are a dull grey and heavily fringed with hairs.

The female moth lays small white eggs singly or in groups of about six. She may lay as many as 350 during her lifetime. In warm conditions, such as the summer months, the eggs will hatch out in about four days, and the larva will mature in about three weeks or less. The larvae can crawl backwards and forwards and do so very actively, leaving a trail of webbing wherever they go. Once the larva is mature, it will leave the food source and find a small crevice where it will spin a white cocoon and change to the pupa state. The pupa stage lasts about one week before the adult moth emerges, although this can take much long in cold conditions.

Granary Weevil

Insects infesting grain are often called weevils but there are only two that are really weevils—the granary weevil (*Sitophilus granarius)* and the rice weevil. They are very similar in appearance and are both widely distributed. The granary weevil at one time in history could probably fly, however it has been infesting homes, mills, and granaries for so long that it has lost the use of its wings and is now strictly an indoor pest.

The adult granary weevil is a shiny reddish-brown colour with a long thin snout, which terminates in minute jaws. It is about 3mm in length and has grooved wing covers and a lightly dented thorax.

The female granary weevil lays her eggs over an extended period; she lays each egg inside a kernel of corn or wheat by chewing a hole into the kernel. When the egg hatches out, the small white larva lives inside the kernel and eats the dry insides. In wheat kernels there is only enough space for one larva however corn kernels can house up to three. During the warm summer months, the complete life cycle of the granary weevil can take place in about six weeks.

In the case of granary weevils, the long-living adult can cause as much damage as the larva. This is unusual for grain eating insects; it is usually the larva that causes all the damage.

Rice Weevil

The rice weevil (*Sitophilus oryzae)* was originally discovered in rice and researchers believe it originated in India. It is not as widely established as the granary weevil, however it causes more damage. They look very similar but the rice weevil is dull brown in colour in contrast to the very shiny brown of the granary weevil. The most distinctive feature of the rice weevil is the four red spots in each corner of the wing covers. Unlike the granary weevil, it is a strong flyer, although it is more likely to be seen indoors because it infests rice, crackers, cake mixes, breads, etc.

The female lays an egg in a kernel of corn or grain of wheat or barley. The egg hatches in about three days and the white, short, fat larva lives inside the kernel for about sixteen days before it changes into a pupa. The pupa remains in the kernel for three to nine days before becoming an adult. It doesn't leave the kernel right away; it stays inside for a few days eating the inside of the kernel. During the summer months the complete life cycle of the rice weevil is about thirty-five days.

SILVERFISH

About the Silverfish

Books and papers are especially prone to silverfish damage. The silverfish's glistening, scaly body, and fish-like shape along with its very agile movements and its habit of quickly scurrying away if discovered, have given rise to such names as silverfish, silver moth, silver louse, sugar fish, etc. This little creature tends to stay away from light, which makes it difficult to capture. It is a particular problem in libraries.

The silverfish is about 8mm long and tapers from the head to the tail end of the body; it belongs to Thysanura, the lowest order of insets. Silverfish have two distinctive antennae and three long hair like appendages along the sides of their bodies, however the entire body is covered in silvery scales. It has six legs that originate at the thorax and are very powerful though they are not very long. These strong legs allow them to run surprisingly fast.

Controlling Silverfish

One of the best ways to control these destructive pests is with natural pyrethrum powder or diatomaceous earth (not pool grade). Sprinkle or puff diatomaceous earth or pyrethrum powder along skirting boards, on bookshelves, into cracks and crevices and in areas where you suspect silverfish activity.

You should begin treating for silverfish as soon as you notice them, as they can become a problem if allowed to go untreated. Once a nest becomes established in your home it is nearly impossible to eliminate them.

You can also apply diatomaceous earth outside along the foundation, under sidings, light fixtures, and electrical boxes. Since silverfish like to make their homes in attics and basements, it is a good idea to sprinkle diatomaceous earth liberally in these places. For best results you should reapply diatomaceous earth about every eight months indoors and every three months outdoors.

You may continue to see a few silverfish wandering about foraging for food even after you have thoroughly treated your home. This is normal, as the dust will take some time to take effect.

If you notice more dead silverfish in one area than others, it is likely they are near a nesting site. Pay close attention to these areas, dust with diatomaceous earth or natural pyrethrum powder and monitor silverfish activity. You may also want to consider placing a few glue traps in areas with high silverfish activity.

SILVERFISH CONTROL RECIPE

SILVERFISH BE GONE SPRAY

This natural recipe will help you combat silverfish activity.

Ingredients:

4 cups natural orange based castile soap
4 drops peppermint essential oil
2 drops neem essential oil

Yields: 4 cups

Time to make: About 5 minutes
Shelf life: Indefinite
Storage: Spray bottle and glass jar with tightly fitting lid

Method:

Combine orange based soap and your essential oils in a glass jar, secure lid, and shake vigorously to blend. Pour half of mixture into a spray bottle and store the other half in the glass jar.

How to use:

Mist along skirting boards, foundations, etc to repel silverfish. Spray along walls, or in areas of known silverfish activity. Re-spray daily or as often as needed. You can also use a paint brush to apply to areas such as crown moulding or the tops of bookshelves.

Note: Be sure to test in an inconspicuous spot before misting on walls, flooring, etc.

Tips and Hints for Silverfish Control

- Keep your indoor home temperature cool. Cool temperatures will help stop silverfish breeding cycles.
- You can prevent silverfish from climbing up bookshelves and walls by creating a natural barrier with chalk.
- Frequently moving the books on your shelves will help reduce silverfish activity in the books. Most silverfish damage occurs in items that have remained undisturbed in moist or humid conditions for a year or more.

SPIDERS

Because Australia is home to more than 2000 species of spiders, including some of the most poisonous in the world, it is a good idea to become familiar with the most common varieties you are likely to encounter. Even though spiders can be dangerous they are also beneficial insects and help control other annoying pests, such as flies, moths, cockroaches, etc. Always use caution and common sense when removing them or treating for them.

Seek immediate medical attention if you are bitten by a spider. If possible, capture the spider that bit and do not try to identify the spider yourself as some very dangerous species look very similar to those which are harmless. It is always best to allow a professional entomologist to identify the species.

The Funnel Web

The male of this species dies soon after mating. The female spider however, can live for eight years or in some cases longer.

The funnel web is perhaps the world's most venomous spider. It is impressive to look at, being 35mm (female) and 25mm (male) in size with large fangs. It is unique to Australia and occasionally makes its home in our gardens. Its name comes from the unique funnel shape of the web it makes.

This spider is very aggressive, especially during mating season when it is much more likely to enter homes. Use extreme caution around funnel web spiders and seek medical attention immediately if bitten.

Mouse Spider

The mouse spider is found in all parts of Australia except Tasmania.

Males tend to wander in the spring and autumn and are much more active than the females, who tend to spend their entire lives in the oval shaped burrow they create. The females are more aggressive and highly poisonous and are equipped with long large fangs.

In case of a mouse spider bite apply pressure and immobilise the area bitten, seek immediate medical attention.

The Redback Spider

The Redback is also one of the most deadly spiders in the world.

The female Redback spider has a red stripe down the middle of her abdomen and is otherwise completely black. She is about 10mm in length. The male has a white abdomen with four black stripes running horizontally and may also have a small red vertical strip. The male is typically about 4mm in length.

White-tailed Spiders

White-tailed spiders are so named because of the distinctive white marking on the tail of the abdomen. Other then the white mark they are a dull brownish-grey with shiny brown legs, and a cylindrical shaped body. The white-tailed spider is about 1-2 cm in length, the females being larger than the males.

Black House Spiders

This species is primarily found in eastern and southern Australia, and is also called the window spider because of its habit of making its messy webs in the corners of windows. The abdomen of the black house spider is dark grey and the legs are black to dark brown. The female is about 18mm, while the male is smaller at about 8mm in length.

Brown Widow Spider

We are more likely to encounter this spider in our homes than those previously discussed. While the brown widow spider is in the same family as the black widow spider and the Redback it is about 1/10 as poisonous. It is also much less likely to bite then its more aggressive relatives. It may appear black at first glance, but it is really a deep mahogany brown. The brown widow spider's egg case has tuffs of silk and a rough texture.

Among those who seek to live in harmony with nature, spiders are considered very beneficial. It is only when the spiders become a pest, or especially a danger, do we consider eradicating them from our homes.

Spider Control

Spiders are beneficial insects that act as a natural pest control for other disease carrying insects such as cockroaches. If you have a spider in an inconvenient place in your home, you should try relocating it. However, dangerous species should not be handled or touched. These species are best killed by misting with a natural spider spray (see spider control recipes).

Here are some other useful tips:

- To prevent spiders from making homes in electrical outlets or vents puff a little diatomaceous earth in these areas.
- If there is an area in your home where dusting or spraying would be difficult or inappropriate, remove the webs and use glue traps. Glue traps let you see what type of spiders, and other insects, you have in your home.

- A small, hand held electric zapper is a safe way of killing aggressive spiders and can also come in handy for killing wasps, flies, bees, etc.
- Vacuuming up a spider and its web up is another solution. Be sure to also vacuum up some diatomaceous earth with the spider to ensure it dies. If you have a bag less vacuum cleaner and you are concerned the spider might be venomous allow the vacuum cleaner to sit outside for a few days before emptying it.
- Attics, basements, and crawl spaces are all areas of concern as they can be quiet, dark, and rarely disturbed places for spiders to make their homes.
- To prevent spiders from invading these areas make sure all holes, cracks, or other places they could enter have been sealed up or screened. If you currently have a spider infestation in one of the locations, treat by puffing diatomaceous earth generally over the area. Diatomaceous earth will also help control other invading insects and the treatment lasts up to one year!

SPIDER CONTROL RECIPE

SPIDER SLAY SPRAY

A natural spider spray formula for inside your home.

Ingredients:

225 grams of any natural peppermint, citrus or lavender soap, shredded or liquid castile soap
1 litre of water
Up to 8 drops of any one, or combination of these essential oils:

- **Citronella**
- **Lavender**
- **Cinnamon**
- **Peppermint**
- **Citrus**
- **Tea tree**
- **Neem**

Yields: 3 cups

Time to make: About 5 minutes
Shelf life: Indefinite
Storage: Spray bottle and glass jar

Method:

Combine water, natural soap, and essential oils in a glass jar, secure lid, and shake vigorously to blend. Pour half of mixture into a spray bottle and store the other half in the glass jar.

How to use:

Before you spray this solution, sweep any spider webs or vacuum them. If you use a broom, treat the bristles outdoors with diatomaceous earth. Carefully inspect the broom bristles for any spider egg sacs and squish or burn any you find.

In the case of non-poisonous spiders you could remove the egg sac to your garden, however do be sure they are a non-venomous kind! If in doubt destroy it.

Mist solution into closets, drawers, pantries, storage containers, or anywhere you have a problem with spiders. It not only kills spiders but it also appears to deter other spiders from making their webs where it has been sprayed.

Note: Any of the above oils will enable this formula to kill spiders and other insects on contact! Test for strength and your favourite scent.

Variation:

You can greatly enhance this formula with this wonderful tea:

Orange peel tea

Take the peelings from 4-6 large oranges, and place in1 litre of boiling water for 5 minutes, then turn off the heat, and allow steeping until cool. Strain the peels and toss them into your compost heap. Use this water whenever a recipe in this book calls for water.

Body Care

In this section, I have included my favourite recipes for natural body insect repellents, shampoos, lotions, and ointments.

INSECT BITES AND STINGS

These natural remedies are particularly soothing for bee, wasp, and hornet stings, as well as bites from flies, mosquitoes, etc.

Lavender

Lavender is wonderful for bringing relief from bites or stings. It eases pain, helps stop itching and reduces the chances of an infection. Place a drop of undiluted lavender essential oil directly on the injured area.

Natural Aloe Vera

Aloe Vera is also very soothing for bites. Apply some to the injured area to sooth itching and reduce pain.

Bicarbonate of Soda

You can make a poultice of bicarbonate soda and vinegar to aid in the relief and healing of bug bites. Place the bicarbonate of soda and vinegar poultice directly on the affected area. Change the vinegar to water to treat a bee sting.

Cornstarch

Cornstarch works well for drawing out poisons from many insect bites, plus it also soothes diaper rash. Combine enough water and cornstarch to make a paste. Apply this salve directly on to diaper rash or insect bites.

Jewel Weed

Works well to ease the itch caused by poison ivy, poison oak, and mosquito bites. Apply jewel weed to affected areas and rub.

Lime

Lime juice works especially well for soothing bug bites. Apply a mixture of lime juice and water with a cotton ball to affected areas.

INSECT REPELLENTS

These natural remedies will help keep the bugs away so you can enjoy the outdoors.

Every person is different and can accept different amounts of essential oils on their skin. As a general rule it is best to dilute the essential oil in a base oil such as jojoba, almond etc instead of applying them straight onto the skin. You will need to experiment to find your favourite scents, the most affective repellents, and the correct amounts for you and your family. The amounts in the recipes are there as a guide only, as a base from which to start experimenting.

Here is a list of herbs and oils found to be particularly useful in repelling insects. Combine your favourites together to create your very own insect repellent.

Herb/Oil (check to see if ok in pregnancy)

- Basil
- Bay
- Bergamot
- Cayenne pepper
- Chamomile
- Chives
- Citronella*
- Citrus
- Clover
- Coriander seeds
- Feverfew
- Lavender
- Lemon balm*

- Neem
- Peppermint
- Pennyroyal**
- Rose geranium
- Rosemary
- Sage**
- Tarragon
- Tea tree oil
- Thyme
- Wormwood**

*attracts bees

** Do not use if pregnant.

INSECT REPELLENT RECIPES

FLORAL INSECT REPELLENT MIST

This lovely floral essential based repellent works especially well against sand flies and mosquitoes.

Ingredients:

- **2 ½ cups of water**
- **2 tbsp aromatherapy carrier oil (this helps to blend the water and essential oils)**
- **10 drops each of citronella, lavender, lemongrass, and pennyroyal essential oil or your favourite oils from the repellent list**

Yields: 2 ½ cups

Time to make: About 5 minutes
Shelf life: Indefinite
Storage: Spray bottle

Method:

Combine water and your essential oils in a spray bottle, secure lid, and shake vigorously to blend.

How to use:

Mist onto your skin before you go outdoors. Avoid eyes and do not mist on your face. To apply to your face apply a little of the insect repelling solution to a cotton ball and carefully dab your face.

Note: Pennyroyal is contraindicated in pregnancy. Consult your health care professional or herbalist before using. Keep out of the reach of children and pets to avoid accidental ingestion.

VINEGAR AND SPICE REPELLENT

This repellent uses natural apple cider vinegar as a base.

Ingredients:

- **½ cup apple cider vinegar**
- **25 drops of your favourite essential oil or oil combination**

Yields: ½ cup

Time to make: About 5 minutes
Shelf life: Indefinite
Storage: Spray bottle

Method:

Combine apple cider vinegar and your essential oils in a spray bottle, secure lid, and shake vigorously to blend.

How to use:

Mist on to exposed body parts to repel insects.

HERBAL OIL INSECT REPELLENT

An oil base will make your repellent last longer.

Ingredients:

2 tablespoons vegetable oil (e.g. almond, olive, jojoba)
15 drops of your favourite essential oil or oil combination

Yields: 2 tablespoons

Time to make: About 5 minutes
Shelf life: Indefinite
Storage: Glass jar or bottle with lid

Method:

Combine vegetable oil and your essential oils in a glass jar, secure lid, and shake vigorously to blend.

How to use:

Using hands or a soft cloth, rub the mixture onto exposed body parts to repel insects.

Gabriela Rosa

LICE

Lice have been a problem for people for thousands of years. Even Aristotle referred to these parasites in his writing.

Lice can't jump and they do not have wings, so they are mainly spread by direct contact, such as wearing an infested person's hat or by sitting next to an infected person. Children's sleep-overs are of particular concern as the children may share hair brushes, sleeping bags, pillows, etc. Once established, lice are difficult to eliminate. It may take several rounds of shampoo and combing before they are completely removed.

Lice are white to grey in colour and about 1.5 to 3mm in length. They can be seen by the unaided eye. The eggs however, which are also called nits, are sometimes difficult to see. They are pear shaped and glued to the base of the hairs by a sticky substance secreted by the female louse. The female louse will lay at least 50 eggs in her life time. Eggs hatch in about 6 days under favourable conditions.

The bite of the louse is usually not painful however the saliva injected at the time of the bite can cause itching and redness in some people.

Head Lice Treatment

Lice can transmit diseases, so it is extremely important to address lice as soon as they are discovered.

Following these steps should ensure complete extermination of head lice.

What you need:

Nit comb (metal ones work better and are easier to clean)

Essential oils:

- Eucalyptus
- Lavender
- Neem
- Rosemary
- Tea tree

Soap:

- Coconut oil castile soap
- Olive oil castile soap

Shampoo:

See Lice Shampoo recipes

Caution: Supervise children when using essential oils in shampoos, and be very careful not to get shampoos containing essential oils in eyes. Always rinse hair away from your face.

Note: Removing lice is a lengthy process and depends on the extent of the infestation, the hair type and length. So, set aside a few hours for the treatment and be patient, it may take several evenings dedicated to this treatment to completely eliminate the lice.

Step 1.

Using a coconut or olive oil based castile soap thoroughly wash hair. As you wash add 3 drops each of tea tree oil and neem essential oil. Rinse hair thoroughly.

Step 2.

Repeat step one, except once you have lathered your hair up a second time, do not rinse, leave in the lather and wrap hair in a towel. Now take a break and let it infuse for about 45 minutes.

Step 3.

After 45 minutes you are ready to comb hair using a fine toothed nit comb. Make sure hair is free from tangles (by using a wide-toothed comb if necessary) before using the nit comb. Comb by starting at the very root of the hair and drawing the comb to the end. Do this strand by stand until all the nits have been removed. You can also use a magnifying glass as you comb to help find the nits. Clean the comb as you go and moisten hair as needed as you work.

Step 4.

Once you are done combing, rinse your hair thoroughly and rewash.

Step 5.

Clean your nit comb by dipping in alcohol. You should also clean bedding, hands, clothing, and hats. You can place these in the freezer overnight to kill any nits or lice or by washing clothing and bedding in hot water.

Step 6.

You may need to repeat this treatment several times before the lice are eliminated. Be sure to run a nit comb through the hair daily until you no longer find any nits.

HEAD LICE TREATMENT RECIPES

These shampoos will help you control and eliminate head lice.

PEPPERMINT SHAMPOO

Not only does this shampoo limit the reproduction of lice, it smells fantastic.

Ingredients:

- **2 drops peppermint essential oil**
- **2 drops orange essential oil**
- **1 bottle of tea tree essential oil based shampoo**

Yields: 1 ½ cups

Time to make: About 5 minutes
Shelf life: Indefinite
Storage: Shampoo bottle

Method:

Add the essential oils directly to the shampoo bottle, secure lid, and shake vigorously to blend.

How to use:

Use as normal shampoo.

CREAMY CONDITIONER

This wonderful creamy mixture loosens the glue around the egg. The glue used by lice to attach their eggs to the hair shaft is stronger than cement. Using this cream as part of a thorough delousing helps to ensure quick and safe removal of eggs.

Ingredients:

- **1 cup olive oil**
- **1 unpeeled cucumber**
- **2 drops rose hip oil**
- **2 -3 ground apricot pits**

Yields: 1 cup

Time to make: About 30 minutes
Shelf life: 1 week in the refrigerator
Storage: Glass jar with tightly fitting lid

Method:

Combine olive oil, unpeeled cucumber, apricot pits, and rose hip oil in a blender dedicated to this purpose, blend until creamy.

How to use:

Use a wide tooth comb to work this all through your hair, and be sure to work well into scalp. Cover with a disposable shower cap for at least an hour, overnight would be best. This will loosen the lice eggs so they fall off easily during combing with the nit comb. Store excess cream in a glass jar in the refrigerator.

LICE PREVENTION CONDITIONER

Having head lice is no fun and can disrupt the entire home, so why not try to prevent it if possible? Lice hate tea tree oil. Apply this cream to your hair once or twice a week during lice season, and hopefully ward off the little beasts before they have a biting chance.

Ingredients:

2 drops tea tree oil
Your regular natural conditioner

Yields: 1 cup

Time to make: About 30 minutes
Shelf life: 1 week in the refrigerator
Storage: Glass jar with tightly fitting lid

Method:

None!

How to use:

Massage the tea tree oil into your hair before your regular hair conditioner, and leave on for 10 minutes before rinsing out. Again, if you leave this on overnight, and rinse it in the morning, all the better. Two drops of tea tree oil is worth a pound of cure!

EUCALYPTUS LAUNDRY BOOSTER

Eucalyptus oil has been used in lice fighting for ages. As a laundry additive, it aids in eliminating lice and nits from your family's bedding and clothes.

As unpleasant as it seems, there is a chance live lice and nits are resting in your bedding and clothing.

Ingredients:

10 drops eucalyptus essential oil (or neem or tea tree essential oils)
Your regular laundry detergent

Yields: 1 regular laundry load

Time to make: About 30 minutes
Shelf life: 1 week in the refrigerator
Storage: Glass jar with tightly fitting lid

Method:

Combine two drops of eucalyptus oil with your regular measure of laundry detergent.

How to use:

Pour the formula into your washing machine as the water fills, prior to loading the clothes.

Note: Drying your clothes, bedding, towels, etc, on high heat in your clothes dryer will also kill lice and nits living in these items.

INDISPENSABLE OILS

The following is a list of oils you will find especially helpful in your battle against lice. Experiment with the scents and strengths you prefer.

Neem Oil

Affectionately called 'the village pharmacy', India's neem tree is a natural first aid kit. Neem extracts have properties that interfere with the life cycle of parasites, inhibit their ability to feed and prevent the eggs from hatching.

Tea Tree Oil

Loaded with the active ingredient terpinen-4-ol, tea tree oil is highly prized for its versatility. Known as our own 'wonder from down under' oil, it comes from the melaleuca tree, which is native to Australia and has long been used by the aborigines for countless purposes.

Karanja Oil

This oil, derived from the karanja or pongam tree, has incredible insecticidal qualities. Very similar to neem oil, it also has antiseptic, anti-parasitic and cleansing properties. It is extremely useful in the treatment of head lice, scabies, general itchiness, herpes, eczema and most sores. A truly wonderful plant, the beautiful karanja has been prized for many centuries in its native India.

Eucalyptus Oil

This oil has strong antiseptic and germicidal properties and can be used as both deodoriser and sanitizer. The oil can be used as an insect repellent and for the relief of insect bites.

Lavender Oil

Lavender oil and tea tree oil can prevent the spread of head lice because the louse is repulsed by the smell of these oils! It is a calming, relaxing oil, so soothes the frustration of the endless combing as well. Besides its wonderful qualities, it smells great!

Peppermint Oil

Peppermint is widely cultivated for its fragrant oil, and it has been used historically for numerous health conditions, including the prevention and treatment of head lice.

Rosemary Oil

Also known as Incensier, rosemary is a crisp and clean smelling essential oil. It is great for stimulating the brain, improving memory and mental clarity. It is also used for improving hair and scalp health as well as repelling insects, including head lice.

Catnip Oil

Known for its intoxicating effects on our feline friends, catnip oil also repels and even kills insects. Catnip has been found to be ten times more effective at repelling mosquitoes than the compound used in most commercial bug repellents!

Erigeron Oil

The effectiveness of erigeron oil is evident from its commonly used nickname 'Fleabane'.

Olive Oil

Olive oil smothers and kills active head lice. You will soon notice how using olive oil makes nit removal easier and re-moisturises your scalp beautifully. Any time you wish to treat your hair to a lovely conditioner, just massage in some olive oil, with a drop of rose hip oil, and leave in an hour, or overnight. You will not believe the texture of your hair, and the lice will curl up and die!

Coconut Oil

Fractionated coconut oil is remarkable! It has a virtually indefinite shelf life, is non-greasy and non-staining. It provides essential fatty acids, which also make it a great moisturiser!

SCABIES MITES

Scabies are parasitic mites, meaning they are nasty little beasts that prey on your flesh and blood.

About the mite

When the female mite finds a human host, they burrow under the skin to feed and lay their eggs. The movement of the mites under the skin causes intense itching and produces a very distinctive, blotchy red rash. The eggs hatch in a week to ten days, mature into nymphs, and then become adult mites soon thereafter.

Adult mites only live in your skin for three to four weeks, but during that time, they have laid enough eggs to keep infections going indefinitely, if left untreated.

Despite popular belief, catching scabies isn't a sign of poor hygiene; it simply means you've had a spot of bad luck. Scabies is easily transmitted through skin-to-skin contact, which means you can contract them in just about any public place. It's also fairly common to catch scabies by sharing clothing, towels, or bed sheets with an infected person.

The scary part is that it takes four to six weeks for your skin to become irritated enough to show symptoms. This means a person could be spreading scabies for over a month before they realise they're infected.

SCABIES MITE TREATMENT RECIPES

This is a topical cream that will calm the itching and soothes your cracked skin. You will find two separate recipes, one for daytime use, and one best used at bedtime. The bedtime cream helps insure against re-infection, so you can sleep easy, assured that you're constantly moving towards complete elimination of this pest.

DAYTIME CREAM

Ingredients:

- **2 drops jewel weed essential oil**
- **2 drops ginger essential oil**
- **2 drops Echinacea essential oil**
- **1 cup coconut oil**

Yields: ½ cup

Time to make: About 10 minutes
Shelf life: Indefinite
Storage: Glass jar with tightly fitting lid

Method:

In a glass jar, blend together essential oils and coconut oil.

How to use:

Use generously, whenever the itching is uncomfortable.

BEDTIME CREAM

Ingredients:

- **2 drops almond essential oil**
- **2 drops tea tree essential oil**
- **2 drops neem essential oil**
- **2 drops jojoba essential oil**
- **½ cup coconut oil**

Yields: ½ cup

Time to make: About 10 minutes
Shelf life: Indefinite
Storage: Glass jar with tightly fitting lid

Method:

In a glass jar, blend together essential oils and coconut oil.

How to use:

Use generously at bedtime, and try to sleep in 100% cotton bed clothes.

TICKS

Even if you haven't been hiking in the bush, you can still pick up ticks in your own backyard, or from pets.

Ticks are members of the same class as spiders, scabies, and scorpions and are actually giant parasitic mites. Ticks mainly live on wild animals however they will attach to people and suck their blood. The sucking part of their mouth, the rostrum, has a hook on the end that they use to anchor themselves under the skin. The male is smaller and less frequently seen than the female. The body of the female expands as she becomes engorged with blood. Once she has fed, she will drop off the host and lay a mass of between 3,000 and 6,000 brownish eggs.

Ticks frequently attack pets. You should check your pet often for ticks and remove any you find. If you find ticks in your garden, you should apply diatomaceous earth in any areas where you suspect tick activity. Ticks carry many dangerous diseases such as babesiosis, lyme, ehrlichiosis. If you discover a tick biting you or your pet, it is important to remove it immediately.

Using tweezers, grasp the tick's head and mouth parts as close to the skin as you can. Tweezers with a fine point will work the best. Pull with steady, gentle pressure, easing the tick out. Never yank or pull roughly or abruptly, as this will case the tick's head to remain under the skin. Be careful not to squish, squeeze, or crush the body of the tick as this can cause it to release more infective fluids into the wound. Ticks implant themselves firmly under the skin so it may take a while to get the barbs and hooks to release.

Even careful pulling can still result in part of the tick being left under the skin. If this happens, don't be alarmed. The mouth parts do not transmit disease and it is just like having a splinter under your skin.

With time, it will be worked out naturally by your body. However, just as with a splinter there is a chance you could get a secondary infection, if you have any concerns please contact your health care professional immediately.

Once you have removed the tick do not throw it away! Place it in a sealable plastic bag or clean dry jar and write the date on the outside. You should keep the tick in case you begin experiencing disease symptoms. The tick will help your physician and medical professionals with diagnosis and treatment. All tick transmitted diseases will appear within one month, so you may throw the tick away after that time.

A popular myth is that holding a lit match to the tick will make it fall off. However, this does not work and is very dangerous, as it can cause burns. Other tick removal suggestions you may have heard might include suffocating the tick using nail polish or petroleum jelly; these seldom work and will most likely cause the tick to release more infectious liquids into the wound.

TICK REPELLENT RECIPES

GARLIC TICK REPELLENT

An effective natural bug repellent.

Ingredients:

1 ¼ cup water
¼ cup garlic juice

Yields: 1 ½ cup

Time to make: About 5 minutes
Shelf life: Indefinite
Storage: Spray bottle

Method:

Combine water and garlic juice in a spray bottle, secure lid, and shake vigorously to blend.

How to use:

Shake well before using. Spray lightly on exposed body parts for an effective repellent lasting up to 5 - 6 hours. Strips of cotton cloth can also be dipped in this mixture and hung in areas, such as patios, as a localised deterrent.

ESSENTIAL EUCALYPTUS REPELLENT

This safe, organic repellent keeps the bugs away.

Ingredients:

2 cups of water
3 teaspoons eucalyptus essential oil

Yields: 2 cups

Time to make: About 5 minutes
Shelf life: Indefinite
Storage: Spray bottle

Method:

Combine water and eucalyptus essential oil in a spray bottle, secure lid, and shake vigorously to blend.

How to use:

Mist any exposed areas of your body with the solution.

Note: For longer-lasting protection, combine the eucalyptus oil with vegetable oil instead of water, store it in a jar or vial, and apply with your hands spreading over all exposed skin.

Pet Care

Pets are an important part of our lives. Whether we have a dog or a cat our pets provide us with caring companions, and, especially in the case of cats, can play an important role in natural pest control.

Nothing is worse than seeing your dog or cat suffering from a flea infestation. While you may be tempted to run out and buy a commercial flea control shampoo or collar, you should think twice, bathing your pet in toxic chemicals doesn't treat them with the care they deserve, especially when there are natural shampoos and remedies which work just as well if not better than their harsh chemical counterparts.

This section is for our furry friends, because just like us they should never be subjected to harsh toxic commercial chemicals and whilst you are in contact with chemicals for treating your pets it may also be a threat to your health and fertility—another important reason to avoid this type of exposure.

NATURAL PET SHAMPOOS & OTHER PET CARE RECIPES

Keeping your pet clean will go a long way towards preventing parasites taking hold. You should brush your pets daily and finish by running a flea comb through their fur. You should also bathe your pet about once every 10-14 days using a natural shampoo—if they are actually suffering a current infestation you may need to increase frequency. However be careful not to overdo it because too much washing can cause skin problems in some pets. The following shampoo recipes will leave your pet clean, fresh, and free from toxic chemicals.

EVERYDAY PET SHAMPOO

This is a great shampoo to use on your pet on bath day.

Ingredients:

- **1½ tablespoons liquid castile soap**
- **1 teaspoon glycerine**
- **1¼ cup water**
- **Optional: 3 drops of your favourite essential oil**

Yields: 1 ½ cups

Time to make: About 10 minutes
Shelf life: Indefinite
Storage: Glass jar with tightly fitting lid

Method:

Combine water, glycerine and essential oil in a glass jar, secure lid, and shake vigorously to blend.

How to use:

Prepare your pet for a bath and wet down the fur, pour shampoo into your hands or directly on your pet's fur and lather. Rinse well, repeat if necessary.

Note: Pets are often sensitive to essential oils, so do not add them unless necessary, such as for flea or tick control.

FIDO'S FAVOURITE DOGGY SHAMPOO

The orange essential oil smells great and will leave your pet with a refreshing scent.

Ingredients:

- **1 teaspoon orange essential oil**
- **1½ tablespoons liquid castile soap**
- **1 teaspoon glycerine**
- **1¼ cup water**

Yields: 1 ½ cups

Time to make: About 10 minutes
Shelf life: Indefinite
Storage: Glass jar with tightly fitting lid

Method:

Combine water, glycerine and orange essential oil in a glass jar, secure lid, and shake vigorously to blend.

How to use:

Prepare your dog's bath and wet down its fur. Pour some shampoo into your hand or directly on your dog's fur and lather. Rinse well, repeat if necessary.

Note: Not suitable for cats as they are averse to citrus essential oils.

DOWN UNDER DOGGY WASH

Eucalyptus essential oil is used in this recipe for great results.

Ingredients:

6 to 10 drops of eucalyptus essential oil
1½ tablespoons liquid castile soap
1 teaspoon glycerine
1¼ cup water

Yields: 1 ½ cups

Time to make: About 10 minutes
Shelf life: Indefinite
Storage: Glass jar with tightly fitting lid

Method:

Combine water, glycerine and eucalyptus essential oil in a glass jar, secure lid, and shake vigorously to blend.

How to use:

Prepare your dog's bath and wet down its fur. Pour some shampoo into your hand or directly on your dog's fur and lather. Rinse well, repeat if necessary.

FLEA CONTROL SHAMPOO

This shampoo will not only leave your pet smelling great but also help you rid your pet and home of fleas.

Ingredients:

To repel fleas:
5 to 8 drops lavender essential oil

To repel ticks:
5 to 8 drops rose geranium essential oil

1½ tablespoons liquid castile soap
1 teaspoon glycerine
1¼ cup water

Yields: 1 ½ cups

Time to make: About 10 minutes
Shelf life: Indefinite
Storage: Glass jar with tightly fitting lid

Method:

Combine water, glycerine and essential oil in a glass jar, secure lid, and shake vigorously to blend.

How to use:

Prepare your pet's bath and wet down its fur. Pour some shampoo into your hand or directly on your pet's fur and lather. Rinse well, repeat if necessary.

REFRESHING RINSE

Vinegar is a great rinse to finish your pet's bath.

Ingredients:

4 cups water
1 ½ cup apple cider vinegar

Yields: 5 ½ cups

Time to make: About 10 minutes
Shelf life: Indefinite
Storage: Glass jar with tightly fitting lid

Method:

Combine vinegar and water in a glass jar, secure lid, and shake vigorously to blend

How to use:

Pour over your pet's fur for a complete bath experience.

Note: Make sure you don't get any of this rinse in your pet's eyes!

DRY DOG ORANGE RUB

Dry skin can make your dog very itchy. This formula uses the natural properties of oranges to moisturise dry skin.

Ingredients:

- **1 cup coconut oil**
- **10 drops of orange essential oil**
- **1¼ teaspoon glycerine**

Yields: 1cup

Time to make: About 10 minutes
Shelf life: Indefinite
Storage: Glass jar with tightly fitting lid, in the fridge

Method:

Place coconut oil over low heat until it becomes liquid, do not boil. Stirring, add orange essential oil and glycerine until it is all blended. Pour liquid into glass jar, wait until it cools and store it in the fridge.

How to use:

Take it out of the fridge half an hour before use and rub onto your pet's itchy spots.

Note: Not suitable for cats as they are averse to citrus essential oils.

BUGS OFF PET MIST

This mist will help keep annoying bugs away from your pets and witch hazel is a great skin tonic.

Ingredients:

- **3 cups witch hazel extract**
- **3 drops citronella essential oil**
- **3 drops eucalyptus essential oil**
- **3 drops pennyroyal essential oil****
- **3 drops tea tree essential oil**

Yields: 3cups

Time to make: About 10 minutes
Shelf life: Indefinite
Storage: Glass jar with tightly fitting lid

Method:

Combine witch hazel extract and essential oils in a glass jar, secure lid, and shake vigorously to blend.

How to use:

Rub about a half teaspoon of this solution into your pet's fur. You can also massage about a half teaspoon into your pet's collar.

Note: Make sure you don't get any of this in your pet's eyes! Just like people, some pets are sensitive to essential oils. Test this recipe first by placing a small dab on your pet's collar. Monitor their reaction. If they seem fine use as recommended.

**Pennyroyal is contraindicated in pregnancy. Avoid use.

CITRUS FLEA WASH

A completely natural flea solution.

Ingredients:

- **1 cup boiling water**
- **1 lemon sliced**

Yields: 1 cup

Time to make: About 10 minutes
Shelf life: Indefinite
Storage: Glass jar with tightly fitting lid

Method:

Carefully score the lemon's rind using a knife to help release its natural oils. Slice the lemon in pieces and place them in a jar. Pour the boiling water over the lemons and steep over night.

How to use:

Wash your dog with this mixture using a sponge or cotton cloth to quickly and effectively kill fleas.

Note: Not suitable for cats as they are averse to citrus essential oils.

ORGANIC FLEA POWDER

This is a quick and very easy to make flea treatment powder for your pets.

Ingredients:

- **1 teaspoon orange essential oil**
- **½ cup bicarbonate of soda**

Yields: ½ cup

Time to make: About 5 minutes
Shelf life: Indefinite
Storage: Glass jar with tightly fitting lid

Method:

Combine bicarbonate of soda and orange essential oil in a glass jar, secure lid, and shake vigorously to blend.

How to use:

Sprinkle this powder onto your pet and massage down to the skin.

Note: Not suitable for cats as they are averse to citrus essential oils.

BEDDING AND OUTDOOR AREAS

Bedding

Your pet's bedding is of particular concern. Fleas' eggs can fall off easily especially when a pet lies down or stands up. Bedding can harbour fleas that have the potential to migrate into other parts of the home, creating a flea infestation. Washing your pet's bedding frequently in hot water with a few drops of eucalyptus essential oil will also greatly help in reducing fleas.

ROSE TICK MIST

Ingredients:

- **1 cup water**
- **1 teaspoon rose geranium essential oil**

Yields: 1 cup

Time to make: About 5 minutes
Shelf life: Indefinite
Storage: Spray bottle

Method:

Combine water and rose geranium essential oil in a spray bottle, secure lid, and shake vigorously to blend.

How to use:

Mist over your pet's bedding. This recipe may also be used for outdoor areas.

Extra Tip:

Crush dry rosemary leaves into a fine powder and sprinkle over your pet's bed—it's a great flea and tick repellent.

PET'S OUTDOOR AREAS

Outdoor areas where pets frequent such as a favourite sunny spot in the garden can become a breeding ground for fleas and ticks. Watch your pet next time they are outside and locate these areas for treatment.

Banana Pest Relief

This banana based recipe is a great outdoor flea repellent.

Ingredients:

2 banana peels
Water to cover

Yields: 1 to 2 cup

Time to make: About 10 minutes
Shelf life: Indefinite
Storage: Spray bottle

Method:

Combine water and banana peels in a blender and blend until smooth. Pour in to a regular spray bottle.

How to use:

To kill fleas, mist any outdoor areas frequented by pets.

HOME MADE TICK REPELLENT COLLAR FOR PETS

This is a fast project you can make at home to protect your furry friend! Dip a thin natural rope in undiluted eucalyptus oil, then wrap the rope in a bandanna and tie it around your pet's neck. Re-dip the rope about twice a week, especially during tick season.

Get Your FREE Bonuses Today!

FREE Fertility Advice from 'The Bringer of Babies'

Leading natural fertility specialist, Gabriela Rosa (aka The Bringer of Babies) has a gift for you. As a thank you for purchasing this book get your FREE "Natural Fertility Booster" subscription and discover...

- Easy ways to comprehensively boost your fertility and conceive naturally, even for women over 40;
- Natural methods to dramatically increase your chances of creating a baby through assisted reproductive technologies such as IUI, IVF, GIFT or ICSI;
- Simple strategies to help you take home a healthier baby;
- How to prevent miscarriages.

You will also receive the FREE audio CD "11 Proven Steps To Create The Pregnancy You Desire And Take Home The Healthy Baby of Your Dreams" a total value of $397!

Claim your bonuses at

www.NaturalFertilityBoost.com

Be quick, this is a limited offer.

(Your free subscription code is: PYF)

Contacts and Resources

The Natural Fertility Solution *Take-Home* Program

A whole person approach is fundamental to effectively restore and optimize natural fertility in both men and women.

Based upon this key observation and many years in private practice Gabriela Rosa developed the comprehensive and healing natural fertility approach she now successfully shares with couples worldwide. The Natural Fertility Solution *Take-Home* Program is based on Gabriela's proven fertility boosting strategies. This program focuses on enabling simultaneous optimization of the key areas (her 11 Pillars of Fertility), which most dependably will help couples to create the baby of their dreams, irrespective of their previous medical history.

Gabriela's approach has proven to be invaluable in helping prospective parents overcome fertility problems; increasing the chances of taking home a healthy baby as well as being able to prevent miscarriages or malformations and abnormalities (even for older couples). Despite its effectiveness in restoring fertility in delicate cases this is not its only application—her method is also an essential toolkit of *paramount importance* for those prospective parents who simply wish to prepare for the healthiest conception and baby—with the intention of giving their child the best possible start in life.

For more information on The Natural Fertility Solution *Take-Home* Program visit **www.BoostYourFertilityNow.com**.

WHAT ELSE IS ON OFFER?

From her Sydney practice, Natural Fertility & Health Solutions Gabriela and her team run The Natural Fertility Solution Program and also offer seminars and workshops as well as one on one and online consultations.

Other programs and modalities offered at Natural Fertility & Health Solutions include:

Programs

- Fertile Emotions Workshop and Whole-Person Fertility Support Groups;
- The PCOS Solution Program;
- The Endo(metriosis) Solution Program;
- Emotional Freedom Technique Workshops;
- Art Therapy;
- Vedic Meditation (transcendental technique);

Modalities

- Acupuncture
- Acupressure
- Chiropractic

- Classical Homoeopathy
- Craniosacral therapy
- Emotional Freedom Techenique
- Herbal Medicines
- Hypnotherapy
- Health & Life Coaching
- Myofascial Release
- Naturopathy
- Nutritional Medicine
- Osteopathy
- Relationship and Individual Counselling

Gabriela is available for interviews and can be contacted via the website www.BoostYourFertilityNow.com or through Natural Fertility & Health Solutions, PoBox 2342, Bondi Junction 1355 | 1300 85 84 90 or +61 2 9369 2655.

OTHER RESOURCES:

100% Pure Essential oils, Bottles, Jars etc

Australia

New Directions Australia

47 Carrington Rd
Marrickville NSW 2204 Australia
T: 612 8577 5999
F: 612 8577 5977
TOLL FREE: 1800 637 697
www.newdirections.com.au

Sydney Essential Oil Co

Unit 4, 2-10 Fountain S
Alexandria NSW 2015 Australia.
+61 2 9565 2828
www.seoc.com.au

Botany Essentials

39 Cranwell St,
Braybrook VIC 3019, Australia
+61 3 9317 9088
www.botanyessentails.com.au

Gabriela Rosa

USA

Wellington Fragrance Co.
33306 Glendale
Livonia, MI 48150

United States Toll Free (800) 411-3593
International (734) 261-5568 / (734) 261-5571 fax

Index

A

About Mice, 136
About the Bedbug, 67
About the Carpet Beetle, 82
About the Clothes Moth, 144
About the Cockroach, 88
About the Common House Fly, 124
About the Dust Mite, 102
About the Flea , 110
About the Mite, 184
About the Mosquito, 131
About the Silverfish, 157
Ant Control Recipes, 55
Ant Jelly, 58
Ant Motel, 55
Ants, 47
Ants Away Mist, 60
Ant Species, 53
Apple Roach Trap, 93
Arctic Ant Mist, 62

B

Baseboards, 74
Basic Ant Control, 49
Basic Bedbug Control, 71
Basil Fly Mist, 129
Bedbug Control Recipes, 77
Bedbugs, 67
Bedding, 202
Bedding and Outdoor Areas, 202
Bedtime Cream, 186
Beetle Eliminator Mist, 86
Bicarbonate of Soda, 167
Black House Spiders , 162
Body Care, 167
Boric Acid, 21
Brown Widow Spider, 163
Bugs at Bay Sachets , 34
Bugs Off Pet Mist, 199

C

Cajun Cockroach Mist, 98
Carpenter Ants, 53
Carpet Beetle Control Powder, 85
Carpet Beetle Control Recipes, 85
Carpet Beetles, 82
Carpeted Floors, 75
Catnip Oil, 182
Chairs and Upholstered Furniture, 75
Citrus Bedbug Mist, 77
Citrus Flea Mist, 119
Citrus Flea Wash, 200
Clothes Moth Control Recipes, 146
Clothes Moths, 144
Cockroach Be-gone, 97
Cockroach Control Recipes, 93
Cockroaches, 88
Cockroach Herbal Mist, 94
Cockroach in Paris, 96
Cockroach Motel, 95
Coconut Oil , 183
Common Myths about Bedbugs, 70
Confused Flour Beetle, 154
Controlling Carpet Beetles, 83
Controlling Clothes Moths, 144
Controlling Cockroaches, 90
Controlling Dust Mites, 103
Controlling Fleas, 111
Controlling Flies, 125
Controlling Mice, 137
Controlling Mosquitoes, 133
Controlling Pantry Pests, 150
Controlling Silverfish, 157
Cornstarch, 168
Creamy Conditioner, 178
Crown Moulding, 74

D

Dark Mealworm, 152
Daytime Cream, 185
Diatomaceous Earth, 19, 37, 79
Disassembling and Cleaning the Bed, 71

Down Under Doggy Wash, 195
Dry Dog Orange Rub, 198
Dust Mite Control Recipes, 105
Dust Mite Mist, 105
Dust Mites, 102

E

Earth Powder, 78
Erigeron Oil, 182
Essential Ant Repellent, 59
Essential Eucalyptus Repellent, 190
Eucalyptus Laundry Booster, 180
Eucalyptus Oil, 181
Eucalyptus Wash Booster, 106
Everyday Pet Shampoo, 193
Extended Control Program, 76

F

Fido's Favourite Doggy Shampoo, 194
Fire Ants, 54
Flea Comb, 113
Flea Control Powder for Carpet and Upholstered Furniture, 118
Flea Control Recipes, 116
Flea Control Shampoo, 196
Fleas, 110
Flies, 124
Floral Insect Repellent Mist, 171
Floral Insect-Repelling Mist, 35
Fly Away Herbal Sachet, 128
Fly No More Paper, 127

G

Garlic Tick Repellent, 189
General Pest Control Recipes, 32
General-purpose Pennyroyal Mist, 41
General-purpose Spice Insecticide, 40
Granary Weevil, 156

H

Havahart traps, 31
Head Lice Treatment, 175
Head Lice Treatment Recipes, 177
Herbal Oil Insect Repellent, 173
Hints and Tips for Bedbug Control, 80
History , 82, 88
History and Habits, 67
Home Care, 30
Home Made Tick Repellent Collar for Pets, 204
Honey Insecticide, 63
House Fly Control Recipes, 127

I

Important Know-How, 27
Indian Meal Moth, 155
Indispensable Oils, 181
Ingredient Glossary, 19
Insect Bites and Stings, 167
Insect Powder, 30
Insect Repellent Recipes, 171
Insect Repellents, 169
In the Pantry Ant Eliminator, 61

J

Jewel Weed, 168

K

Karanja Oil, 181

L

Lavender, 167
Lavender Oil, 182
Lice, 174
Lice Prevention Conditioner , 179
Lime, 168
Liquid Castile Soap (Vegetable oil or glycerine-based), 22

M

Make the 'Flea's Sneeze' , 116
Mediterranean Flour Moth, 153
Mice, 136
Mice Control Recipe, 142
Mice Mist, 142
Minty Flea Mist, 120
Mosquitoes, 131
Moth Away Sachets, 146
Moths and Beetles Found in Food, 150
Mouse Spider, 162

N

Natural Aloe Vera , 167
Natural Citrus Mist, 64
Natural Enemies of the House Fly, 125
Natural Pet Shampoos & Other Pet Care Recipes, 192
Neem Oil, 181

O

Olive Oil, 183
Onion Flea Mist, 122
Organic Essential Oil Insect Repellent, 33
Organic Flea Powder, 201
Organic Home Fumigator, 36
Other Resources:, 211

P

Pennyroyal Flea Mist, 121
Peppermint Oil, 182
Peppermint Shampoo, 177
Pest-Specific Solutions, 47
Pet Care, 191
Pet's Outdoor Areas, 204
Pharaoh Ants, 54
Pheromone Baits and Traps, 30
Plant Derivatives and Extracts, 19
Plants for Your Home and Patio that Prevent Mosquitoes, 134
Pleasing Peppermint Soap Mist, 39

P

Primary Ant Prevention, 49

Q

Queen's Bane Powder, 56
Quick Tips for Ant Control, 65

R

Reassembling and Finishing touches for the Bed, 73
Refreshing Rinse, 197
Reliable Cockroach Destroyer, 99
Rice Weevil, 156
Rosemary Oil, 182
Rose Tick Mist, 203

S

Sawtoothed Grain-Beetle, 153
Scabies Mites, 184
Scabies Mite Treatment Recipes, 185
Silverfish, 157
Silverfish Be Gone Spray, 159
Silverfish Control Recipe, 159
Something Sticky Insect Trap, 42
Spider Control, 163
Spider Control Recipe, 165
Spiders, 161
Spider Slay Spray, 165
Steam Cleaning, 104
Sweet as Honey Insect Trap, 42
Switch plates, electrical outlets, 74
Symptoms of Pesticide Exposure, 3

T

Tea Tree Oil, 181
The Bed, 71, 104
The Bed Frame, 72
The Funnel Web, 161
The Mosquito Bite, 132
The Pest Control Power of Plants, 44

The Redback Spider, 162
The Wonderful Mint Family, 45
Three-Step Ant Control Program, 50
Three Step Flea Control Program, 112
Three Step Roach Control Program, 91
Tick Repellent Recipes, 189
Ticks, 187
Tips and Hints for Clothes Moth Control, 149
Tips and Hints for Controlling Carpet Beetles, 87
Tips and Hints for Controlling Cockroaches, 100
Tips and Hints for Controlling Dust Mites, 108
Tips and Hints for Flea Control, 112
Tips and Hints for House Fly Control, 129
Tips and Hints for Mice and Rodent Control, 143
Tips and Hints for Mosquito Control, 135
Tips and Hints for Silverfish Control, 160
Tips for Cockroach Prevention, 90
Tips for Mice Prevention, 138
Treating Carpets, 83
Treating Closets, 84
Treating Upholstered Furniture, 84
Treatment Program, 104
Treatment Program for Carpet Beetles, 83
Two Step Rodent Control Program, 139

U

Ultrasonic Pest Control, 31
Under the Bed Mist, 107

V

Vacuum, 104
Vinegar and Spice Repellent, 172
Vintage Moth Repellent, 148

W

Washing Soda, 23
What is Natural Pest Control?, 7
When to Call a Professional, 5
White-tailed Spiders, 162
Wooden Floors, 75
Wooden Furniture, 74

Y

Yellow Mealworm, 152